lectures on
quantum mechanics
**perturbed
evolution**

lectures on
quantum mechanics
perturbed
evolution

Berthold-Georg Englert
National University of Singapore, Singapore

World Scientific

NEW JERSEY · LONDON · SINGAPORE · BEIJING · SHANGHAI · HONG KONG · TAIPEI · CHENNAI

Published by

World Scientific Publishing Co. Pte. Ltd.

5 Toh Tuck Link, Singapore 596224

USA office: 27 Warren Street, Suite 401-402, Hackensack, NJ 07601

UK office: 57 Shelton Street, Covent Garden, London WC2H 9HE

British Library Cataloguing-in-Publication Data
A catalogue record for this book is available from the British Library.

LECTURES ON QUANTUM MECHANICS
(In 3 Volumes)
Volume 3: Perturbed Evolution

ISBN-13 978-981-256-790-1 (Set)
ISBN-10 981-256-790-9   (Set)
ISBN-13 978-981-256-791-8 (pbk) (Set)
ISBN-10 981-256-791-7   (pbk) (Set)

ISBN-13 978-981-256-974-5 (Vol. 3)
ISBN-10 981-256-974-X   (Vol. 3)
ISBN-13 978-981-256-975-2 (pbk) (Vol. 3)
ISBN-10 981-256-975-8   (pbk) (Vol. 3)

Printed in Singapore

To my teachers, colleagues, and students

Preface

This book on the *Perturbed Evolution* of quantum systems grew out of a set of lecture notes for a fourth-year undergraduate course at the National University of Singapore (NUS). The reader is expected to be familiar with the subject matter of a solid introduction to quantum mechanics, such as Dirac's formalism of kets and bras, Schrödinger's and Heisenberg's equations of motion, and the standard examples that can be treated exactly, with harmonic oscillators and hydrogen-like atoms among them.

After brief reviews of quantum kinematics and dynamics, including discussions of Bohr's principle of complementarity and Schwinger's quantum action principle, the attention turns to the elements of time-dependent perturbation theory and then to the scattering by localized interactions. Fermi's golden rule, the Born series, and the Lippmann–Schwinger equation are returning themes.

A chapter on general angular momentum prepares the ground for a discussion of indistinguishable particles. The scattering of two particles of the same kind, the basic properties of two-electron atoms, and a glimpse at many-electron atoms illustrate the matter. Throughout the text, the learning student will benefit from the dozens of exercises on the way and the detailed exposition that does not skip intermediate steps.

Two companion books on *Basic Matters* and *Simple Systems* cover the material of the preceding courses at NUS for second- and third-year students, respectively. The three books are, however, not strictly sequential but rather independent of each other and largely self-contained. In fact, there is quite some overlap and a considerable amount of repeated material. While the repetitions send a useful message to the self-studying reader about what is more important and what is less, one could do without them and teach most of *Basic Matters*, *Simple Systems*, and *Perturbed Evolution* in a coherent two-semester course on quantum mechanics.

All three books owe their existence to the outstanding teachers, colleagues, and students from whom I learned so much. I dedicate these lectures to them.

I am grateful for the encouragement of Professors Choo Hiap Oh and Kok Khoo Phua who initiated this project. The professional help by the staff of World Scientific Publishing Co. was crucial for the completion; I acknowledge the invaluable support of Miss Ying Oi Chiew and Miss Lai Fun Kwong with particular gratitude. But nothing would have come about, were it not for the initiative and devotion of Miss Jia Li Goh who turned the original handwritten notes into electronic files that I could then edit.

I wish to thank my dear wife Ola for her continuing understanding and patience by which she is giving me the peace of mind that is the source of all achievements.

Singapore, March 2006 *BG Englert*

Contents

Chapter 1

Basics of Kinematics and Dynamics

1.1 Brief review of basic kinematics

In quantum mechanics, the physical quantities are symbolized by linear operators A, B, ... that act on vectors — elements of a vector space, that is, not physical vectors in the three-dimensional space of our experience. We often speak of *observables* when referring to these linear operators, which is a sloppy use of terminology because, more precisely, the operators are the mathematical symbols that represent the physical "observable" properties, or simply "observables". The vectors they act on come in two kinds: ket vectors $|\dots\rangle$ and bra vectors $\langle\dots|$ (or right vectors and left vectors). The mathematical operation of hermitian conjugation, or as the physicists say: "taking the adjoint", relates them to each other,

$$\langle\dots|^\dagger = |\dots\rangle\,, \quad |\dots\rangle^\dagger = \langle\dots|\,, \tag{1.1.1}$$

where it is understood that the ellipses indicate identical sets of quantum numbers, which serve as the labels that identify the kets and bras.

A measurement of an observable A yields one of the possible measurement results $a_1, a_2, a_3, \dots$, which are complex numbers in general. If it is known that a measurement of A will surely return the value a_j, then we say that the quantum mechanical system is in the state $|a_j\rangle$,

$$|a_j\rangle: \quad A|a_j\rangle = |a_j\rangle a_j\,, \tag{1.1.2}$$

which — mathematically speaking — is an eigenvector equation, here: an eigenket equation. There is also the corresponding eigenbra equation,

$$\langle a_j|A = a_j\langle a_j|\,. \tag{1.1.3}$$

1

The measurement results a_j is the eigenvalue of A in both the eigenket equation (1.1.2) and the eigenbra equation (1.1.3).

Under these circumstances, namely: the system is in state $|a_j\rangle$, the probability of finding the value b_k upon measuring observable B is given by

$$\text{prob}(a_j \to b_k) = \left| \langle b_k | a_j \rangle \right|^2 . \qquad (1.1.4)$$

The complex number $\langle b_k | a_j \rangle$ is the *probability amplitude* to measurement result b_k in state $|a_j\rangle$; its absolute square is the associated probability. This amplitude has all properties that are required of an inner product, in particular

$$
\begin{aligned}
|a\rangle = |a'\rangle + |a''\rangle : & \quad \langle b|a\rangle = \langle b|a'\rangle + \langle b|a''\rangle , \\
|a\rangle = |\alpha\rangle \lambda : & \quad \langle b|a\rangle = \langle b|\alpha\rangle \lambda ,
\end{aligned}
\qquad (1.1.5)
$$

where λ is any complex number, and

$$
\begin{aligned}
&\langle a|b\rangle = \langle b|a\rangle^* , \\
&\langle a|a\rangle \geq 0 \quad \text{with “=” only if } |a\rangle = 0 .
\end{aligned}
\qquad (1.1.6)
$$

In mathematical terms, these properties characterize the kets as elements of an inner-product space or *Hilbert space* (David Hilbert). There is a Hilbert space for the bras as well, related to that of the kets by hermitian conjugation.

The mathematical property $\langle a|b\rangle = \langle b|a\rangle^*$ has a very important physical implication, namely the statement that the two probabilities $\text{prob}(a \to b)$ and $\text{prob}(b \to a)$ are equal,

$$\text{prob}(a_j \to b_k) = \text{prob}(b_k \to a_j) . \qquad (1.1.7)$$

The probabilities for these related, yet different physical processes,

on the left: probability of finding b_k if a_j is the case,

on the right: probability of finding a_j if b_k is the case,

are therefore always equal. There is, of course, a lot of circumstantial evidence for the validity of this fundamental symmetry, but — elementary situations aside — there does not seem to be a systematic direct experimental test.

Different measurement results for the same quantity A exclude each other. This physical fact is expressed by the mathematical statement of orthogonality,

$$\langle a_j | a_k \rangle = 0 \quad \text{if} \quad a_j \neq a_k, \quad \text{or} \quad j \neq k. \tag{1.1.8}$$

Inasmuch as $\text{prob}(a_j \to a_j)$ is the probability that a control measurement confirms what is known, we must have

$$\text{prob}(a_j \to a_j) = 1, \tag{1.1.9}$$

so that $\langle a_j | a_j \rangle = 1$ must hold. Thus

$$\langle a_j | a_k \rangle = \left\{ \begin{matrix} 0 & \text{if} & j \neq k \\ 1 & \text{if} & j = k \end{matrix} \right\} = \delta_{jk}, \tag{1.1.10}$$

where we employ Leopold Kronecker's δ symbol for a compact presentation of this statement of orthonormality.

Each measurement has a result. This physical fact has a mathematical analog as well, namely the completeness relation

$$\sum_j |a_j\rangle\langle a_j| = 1 \quad (= \text{identity operator}). \tag{1.1.11}$$

As an immediate consequence we note that the eigenket equation

$$A|a_j\rangle = |a_j\rangle a_j, \tag{1.1.12}$$

multiplied by $\langle a_j|$ on the right, and then summed over j, yields

$$A = \sum_j |a_j\rangle a_j \langle a_j|, \tag{1.1.13}$$

the so-called spectral decomposition of A. We get the spectral decomposition of $A^\dagger$,

$$A^\dagger = \sum_j |a_j\rangle a_j^* \langle a_j|, \tag{1.1.14}$$

by making use of the familiar product rule for the adjoint,

$$\left(|1\rangle\lambda\langle 2| \right)^\dagger = |2\rangle\lambda^*\langle 1| \tag{1.1.15}$$

for any ket $|1\rangle$, complex number λ, and bra $\langle 2|$.

An apparatus that measures the physical property A, in fact measures all functions of A,

$$f(A)|a_j\rangle = |a_j\rangle f(a_j), \tag{1.1.16}$$

because you just evaluate the function $f(a_j)$ after finding the value a_j. Put differently, it is our free choice whether we want to call the result a_j or $f(a_j)$ when the jth outcome is found. It follows that the spectral decomposition of $f(A)$ is given by

$$f(A) = \sum_j |a_j\rangle f(a_j)\langle a_j| \,. \tag{1.1.17}$$

It makes consistent sense to regard $f(A)$ thus defined as an operator-valued function of operator A. For example, consider the simple function A^2,

$$
\begin{aligned}
f(A) = A^2 &= \left(\sum_j |a_j\rangle a_j \langle a_j| \right)^2 \\
&= \sum_{j,k} (a_j) a_j \underbrace{\langle a_j|a_k\rangle}_{=\,\delta_{jk}} a_k \langle a_k| \\
&= \sum_j |a_j\rangle a_j^2 \langle a_j| = \sum_j |a_j\rangle f(a_j)\langle a_j| \,, \tag{1.1.18}
\end{aligned}
$$

indeed. Similarly, you easily show that it works for other powers of A, then for all polynomials, then for all functions that can be approximated by, or related to, polynomials, and so forth. But what is really needed to ensure that $f(A)$ is well defined, is that the numerical function $f(a_j)$ is well defined for all eigenvalues a_j. As a consequence, two functions of A are the same if they agree for all a_j:

$$f(A) = g(A) \quad \text{if} \quad f(a_j) = g(a_j) \quad \text{for all } j \,. \tag{1.1.19}$$

As an example consider the following exercise.

1-1 Operator A has eigenvalues 0, $+1$, and -1. Write

$$f(A) = e^{i\frac{2\pi}{3}A}$$

as a polynomial in A,

$$f(A) = c_0 + c_1 A + c_2 A^2 \,,$$

with numerical coefficients c_0, c_1, c_2, after first showing that such a polynomial is the most general function of A.

We recall that operators of two particular kinds play special roles in quantum mechanics. These are the *hermitian* operators, named in honor of Charles Hermite, which are equal to their adjoints,

$$\text{hermitian:} \quad H = H^\dagger, \tag{1.1.20}$$

and the *unitary* operators,

$$\text{unitary:} \quad U = \left(U^\dagger\right)^{-1}, \quad UU^\dagger = 1 = U^\dagger U, \tag{1.1.21}$$

for which the inverse equals the adjoint.

1-2 Consider the spectral decomposition (1.1.17) and show that $f(a_j)$ is real if $f(A)$ is hermitian, and that $|f(a_j)| = 1$ if $f(A)$ is unitary. That is: all eigenvalues of a hermitian operator are real; all eigenvalues of a unitary operator are phase factors.

Several observables A, B, C, ... have their state kets $|a_j\rangle$, $|b_k\rangle$, $|c_l\rangle$, ... with probability amplitudes $\langle a_j|b_k\rangle$, $\langle b_k|c_l\rangle$, $\langle c_l|a_j\rangle$, ... These amplitudes are not independent, however, but must obey the composition law

$$\langle a_j|b_k\rangle = \sum_l \langle a_j|c_l\rangle\langle c_l|b_k\rangle, \tag{1.1.22}$$

which we recognize to be a consequence of the completeness of the $|c_l\rangle$ kets. The self-suggesting interpretation

"First there is $|b_k\rangle$, eventually $|a_j\rangle$, and in between $|c_l\rangle$, but we do not know which C value was actually the case and so we must sum over all c_l."

is *wrong*. The assumption of an actual C value at an intermediate stage leads to logical contradictions.

There are two main reasons for this. First, the l sum is not a sum of probabilities but of probability amplitudes. The resulting statement about probabilities reads

$$\text{prob}(b_k \to a_j) = \sum_l \text{prob}(b_k \to c_l)\,\text{prob}(c_l \to a_j)$$
$$+ \sum_{l \neq l'} \langle c_l|b_k\rangle\langle b_k|c_{l'}\rangle\langle c_{l'}|a_j\rangle\langle a_j|c_l\rangle \tag{1.1.23}$$

where the appearance of the $l \neq l'$ terms signifies the possible occurrence of quantum mechanical interferences. Only if the $l \neq l'$ sum happens to vanish, the interpretation above is justified.

Second, there is the fundamental aspect that some observables exclude each other mutually. This feature of quantum mechanics has no true analog in classical physics. In particular, there are pairs of complementary observables. The pair A, B is complementary if the probabilities

$$\text{prob}(a_j \to b_k) = \left| \langle b_k | a_j \rangle \right|^2 \tag{1.1.24}$$

do not depend on the quantum numbers a_j and b_k. Physically speaking: If the system is prepared in a state in which the value of A is known, that is: we can predict with certainty the outcome of a measurement of property A, then all measurement results are equally probable in a measurement of B, and vice versa.

Now, if D were the complementary partner of C, we would have

$$\begin{aligned}
\langle a_j | b_k \rangle &= \sum_l \langle a_j | c_l \rangle \langle c_l | b_k \rangle \\
&= \sum_m \langle a_j | d_m \rangle \langle d_m | b_k \rangle.
\end{aligned} \tag{1.1.25}$$

The wrong interpretation after (1.1.22) would then imply that both C and D have definite, though unknown, values at the intermediate stage, because the two sums are on equal footing. But this is utterly impossible.

Given operator A with its (nondegenerate) eigenvalues a_j and the kets $|a_j\rangle$, can we always find another observable, B, such that A, B are a pair of complementary observables? Yes, we can by an explicit construction, for which

$$|b_k\rangle = \frac{1}{\sqrt{N}} \sum_{j=1}^{N} |a_j\rangle \, e^{i\frac{2\pi}{N} jk} \tag{1.1.26}$$

is the basic example. (More about this shortly.) It is here assumed that we deal with a quantum degree of freedom for which there can be at most N different values for any measurement.

We need to verify that the B states of this construction are orthonormal,

$$\begin{aligned}
\langle b_k | b_l \rangle &= \frac{1}{N} \sum_{j,m} e^{-i\frac{2\pi}{N} jk} \underbrace{\langle a_j | a_m \rangle}_{= \delta_{jm}} e^{i\frac{2\pi}{N} lm} \\
&= \frac{1}{N} \sum_{j=1}^{N} e^{-i\frac{2\pi}{N} j(k-l)} \\
&= \delta_{kl}, \quad \text{indeed.}
\end{aligned} \tag{1.1.27}$$

Then

$$B = \sum_k |b_k\rangle b_k \langle b_k| \tag{1.1.28}$$

with any convenient choice for the nondegenerate B values b_k will do. By construction, we have

$$\left| \langle a_j | b_k \rangle \right|^2 = \left| \frac{1}{\sqrt{N}} e^{i\frac{2\pi}{N}jk} \right|^2 = \frac{1}{N} \tag{1.1.29}$$

so that A, B are a complementary pair, indeed. We note that this property is actually primarily a property of the two bases of kets (and bras) associated with the pair of observables. A common terminology is to call such pairs of bases "mutually unbiased".

In passing, it is worth mentioning that there are quite basic questions about such mutually unbiased basis pairs that do not have a known answer. Quantum kinematics is not a closed subject but still the object of research despite the profound understanding that has resulted from eight decades of intense studies.

1.2 Bohr's principle of complementarity

1.2.1 *Complementary observables*

We consider the situation where we can have at most N different outcomes of a measurement, that is there are no more than N mutually orthogonal states available. One such set is composed of all the eigenstates of some observable A, with the respective kets denoted by $|a_1\rangle$, $|a_2\rangle$, ..., $|a_N\rangle$. Another set is obtained immediately by a cyclic permutation, effected by the unitary operator U,

$$|a_1\rangle \longrightarrow |a_2\rangle = U|a_1\rangle,$$
$$|a_2\rangle \longrightarrow |a_3\rangle = U|a_2\rangle,$$
$$\vdots$$
$$|a_N\rangle \longrightarrow |a_1\rangle = U|a_N\rangle, \tag{1.2.1}$$

generally

$$U|a_j\rangle = |a_{j+1}\rangle, \tag{1.2.2}$$

where the index is to be understood modulo N, so that $|a_{N+1}\rangle = |a_1\rangle$, for example. Applying U twice shifts the index by 2,

$$U^2|a_j\rangle = |a_{j+2}\rangle, \tag{1.2.3}$$

and N such shifts amount to doing nothing,

$$U^N|a_j\rangle = |a_{j+N}\rangle = |a_j\rangle. \tag{1.2.4}$$

Accordingly, we have

$$U^N = 1 \tag{1.2.5}$$

so that U is a unitary operator of period N.

The eigenvalues of U must obey the same equation

$$u^N = 1 \quad \text{if} \quad U|u\rangle = |u\rangle u \tag{1.2.6}$$

for which

$$u_k = e^{i\frac{2\pi}{N}k}, \quad k = 1, 2, \ldots, N \tag{1.2.7}$$

are the possible solutions, all of which occur. We can, therefore, write the equation for U also in the factorized form

$$U^N - 1 = (U - u_1)(U - u_2) \cdots (U - u_N)$$
$$= \prod_{k=1}^{N} (U - u_k). \tag{1.2.8}$$

Let us isolate one factor,

$$U^N - 1 = (U - u_k) \prod_{l(\neq k)} (U - u_l), \tag{1.2.9}$$

and note the following

$$\prod_{l(\neq k)} (U - u_l)|u_m\rangle = \begin{cases} 0 & \text{if } m \neq k, \\ |u_k\rangle\alpha & \text{if } m = k, \end{cases} \tag{1.2.10}$$

with some complex number $\alpha \neq 0$, because one of the factors $U - u_l \to u_m - u_l$ vanishes if $m \neq k$ but all are nonzero if $m = k$. We conclude that the operator acting on $|u_m\rangle$ in (1.2.10) is a numerical multiple of $|u_k\rangle\langle u_k|$, the projector on the kth eigenstate. This product of $N-1$ factors is a polynomial in U of degree $N - 1$, for which we can also give another construction. We apply the familiar identity

$$X^N - 1 = (X - 1)\left(1 + X + X^2 + \cdots + X^{N-1}\right)$$
$$= (X - 1) \sum_{l=0}^{N-1} X^l \qquad (1.2.11)$$

to $X = U/u_k$:

$$U^N - 1 = (U/u_k)^N - 1 = (U/u_k - 1) \sum_{l=0}^{N-1} (U/u_k)^l$$
$$= (U/u_k - 1) \sum_{l=1}^{N} (U/u_k)^l \qquad (1.2.12)$$

where the first step exploits $u_k^N = 1$ and the last step makes use of $(U/u_k)^0 = 1 = (U/u_k)^N$. Now, for $U \to u_k$ the sum equals N, and so we arrive at

$$\left|u_k\right\rangle\left\langle u_k\right| = \frac{1}{N} \sum_{l=1}^{N} (U/u_k)^l . \qquad (1.2.13)$$

1-3 Verify explicitly that

$$\frac{1}{N} \sum_{l=1}^{N} (U/u_k)^l \left|u_m\right\rangle = \left|u_k\right\rangle \delta_{km} .$$

1-4 Verify directly that $\sum_{l=1}^{N} (U/u_k)^l$ is hermitian.

So we know the eigenvalues of U and have an explicit construction for the projectors on those eigenvalues as a function of U itself, and now we find out how the eigenkets of U are related to the original set of kets $\left|a_j\right\rangle$. We begin with

$$\left|u_k\right\rangle\left\langle u_k\middle|a_N\right\rangle = \frac{1}{N} \sum_{l=1}^{N} u_k^{-l} \underbrace{U^l\left|a_N\right\rangle}_{= \left|a_l\right\rangle} \qquad (1.2.14)$$

and then apply $\langle a_N|$ from the left,

$$\langle a_N|u_k\rangle\langle u_k|a_N\rangle = \left|\langle u_k|a_N\rangle\right|^2 = \frac{1}{N}u_k^{-N} = \frac{1}{N}. \qquad (1.2.15)$$

We make use of the freedom to choose the overall complex phase of $|u_k\rangle$ and agree on

$$\langle u_k|a_N\rangle = \frac{1}{\sqrt{N}}, \qquad (1.2.16)$$

with the consequence

$$\begin{aligned}
|u_k\rangle &= \frac{1}{\sqrt{N}}\sum_{l=1}^{N}|a_l\rangle u_k^{-l} \\
&= \frac{1}{\sqrt{N}}\sum_{l=1}^{N}|a_l\rangle\,\mathrm{e}^{-\mathrm{i}\frac{2\pi}{N}kl}
\end{aligned} \qquad (1.2.17)$$

and, after taking the adjoint,

$$\langle u_k| = \frac{1}{\sqrt{N}}\sum_{l=1}^{N}\mathrm{e}^{\mathrm{i}\frac{2\pi}{N}kl}\langle a_l|. \qquad (1.2.18)$$

We read off that

$$\langle u_k|a_l\rangle = \frac{1}{\sqrt{N}}\mathrm{e}^{\mathrm{i}\frac{2\pi}{N}kl}, \qquad (1.2.19)$$

of which the $l = N$ case is (1.2.16).

We have now a second set of bras and kets, for which we can repeat the story of cyclic permutations, effected by the unitary operator V,

$$\begin{aligned}
\langle u_k|V &= \langle u_{k+1}|, \\
\langle u_k|V^2 &= \langle u_{k+2}|, \\
&\ \ \vdots \\
\langle u_k|V^N &= \langle u_k|.
\end{aligned} \qquad (1.2.20)$$

In full analogy with what we did above for U, we conclude here that

$$V^N = 1: \quad V \text{ is unitary with period } N, \qquad (1.2.21)$$

that the eigenvalues of V are $v_l = e^{i\frac{2\pi}{N}l}$, that the projector on the lth eigenvalue is

$$|v_l\rangle\langle v_l| = \frac{1}{N}\sum_{k=1}^{N}(V/v_l)^k, \qquad (1.2.22)$$

and are led to

$$\langle u_N|v_l\rangle\langle v_l| = \frac{1}{N}\sum_{k=1}^{N}v_l^{-k}\langle u_k| \qquad (1.2.23)$$

and then

$$\langle u_N|v_l\rangle\langle v_l|u_N\rangle = |\langle u_N|v_l\rangle|^2 = \frac{1}{N}. \qquad (1.2.24)$$

Here, too, we choose $\langle u_N|v_l\rangle = \frac{1}{\sqrt{N}}$ and establish

$$\langle v_l| = \frac{1}{\sqrt{N}}\sum_{k=1}^{N}e^{-i\frac{2\pi}{N}kl}\langle u_k| \qquad (1.2.25)$$

as well as

$$|v_l\rangle = \frac{1}{\sqrt{N}}\sum_{k=1}^{N}|u_k\rangle e^{i\frac{2\pi}{N}kl}. \qquad (1.2.26)$$

Can we continue like this and get more and more sets of kets? No! Because the kets $|v_l\rangle$ are identical with the kets $|a_l\rangle$, see

$$|v_l\rangle = \sum_{k=1}^{N}|u_k\rangle\underbrace{\frac{1}{\sqrt{N}}e^{i\frac{2\pi}{N}kl}}_{=\langle u_k|a_l\rangle}$$

$$= \underbrace{\left(\sum_{k}|u_k\rangle\langle u_k|\right)}_{=1}|a_l\rangle = |a_l\rangle. \qquad (1.2.27)$$

We have been led back to the initial set of kets.

In summary, we have a pair of reciprocally defined unitary operators,

$$U|v_l\rangle = |v_{l+1}\rangle, \quad \langle u_k|V = \langle u_{k+1}|, \qquad (1.2.28)$$

which are of period N,

$$U^N = 1, \quad V^N = 1. \tag{1.2.29}$$

Their eigenstates are related to each other by the probability amplitudes

$$\langle u_k | v_l \rangle = \frac{1}{\sqrt{N}}\, \mathrm{e}^{\mathrm{i}\frac{2\pi}{N}kl}, \tag{1.2.30}$$

so that the probabilities

$$\left|\langle u_k | v_l \rangle\right|^2 = \frac{1}{N} \tag{1.2.31}$$

do not depend on k and l, which tells us that U and V are a pair of complementary observables.

Being complementary partners of each other, U and V should have a simple commutation relation. We find it by considering the effect of UV and VU upon $\langle u_k |$,

$$\langle u_k | UV = u_k \langle u_k | V = u_k \langle u_{k+1} |,$$
$$\langle u_k | VU = \langle u_{k+1} | U = u_{k+1} \langle u_{k+1} |, \tag{1.2.32}$$

so that

$$\langle u_k | VU = \mathrm{e}^{\mathrm{i}\frac{2\pi}{N}} u_k \langle u_{k+1} | = \mathrm{e}^{\mathrm{i}\frac{2\pi}{N}} \langle u_k | UV \tag{1.2.33}$$

since

$$u_{k+1} = u_k\, \mathrm{e}^{\mathrm{i}\frac{2\pi}{N}}. \tag{1.2.34}$$

The completeness of the bras $\langle u_k |$ now implies

$$UV = \mathrm{e}^{-\mathrm{i}\frac{2\pi}{N}} VU,$$
$$VU = \mathrm{e}^{\mathrm{i}\frac{2\pi}{N}} UV, \tag{1.2.35}$$

and their generalization to

$$U^k V^l = \mathrm{e}^{-\mathrm{i}\frac{2\pi}{N}kl} V^l U^k,$$
$$V^l U^k = \mathrm{e}^{\mathrm{i}\frac{2\pi}{N}kl} U^k V^l \tag{1.2.36}$$

is immediate. These are the *Weyl commutation relations* for the complementary pair U, V, named in honor of Hermann K. H. Weyl.

1.2.2 Algebraic completeness

Now, all functions of U are polynomials of degree $N - 1$, and all functions of V are also such polynomials. Therefore, a general function of both U and V is always of the form

$$f(U, V) = \sum_{k,l=0}^{N-1} f_{kl} U^k V^l = \sum_{k,l=1}^{N} f_{kl} U^k V^l \qquad (1.2.37)$$

or can be brought into this form. It is written here such that all Us are to the left of all Vs in the products, but this is no restriction because the relations (1.2.36) state that other products can always be brought into this U, V-ordered form.

In fact, all such functions of U and V make up all operators for this degree of freedom, which is to say that the complementary pair U, V is *algebraically complete*. To make this point, we consider an arbitrary operator F, and note that then the numbers $\langle u_k | F | v_l \rangle$ are known. We normalize these mixed matrix elements by dividing by $\langle u_k | v_l \rangle$, thus defining the set of N^2 numbers

$$f(u_k, v_l) = \frac{\langle u_k | F | v_l \rangle}{\langle u_k | v_l \rangle}. \qquad (1.2.38)$$

Multiply by $|u_k\rangle\langle u_k|$ from the left, by $|v_l\rangle\langle v_l|$ from the right, and sum over k and l,

$$\begin{aligned}
\sum_{k,l} |u_k\rangle\langle u_k| f(u_k, v_l) |v_l\rangle\langle v_l| &= \sum_{k,l} |u_k\rangle \underbrace{\langle u_k | v_l \rangle f(u_k, v_l)}_{= \langle u_k | F | v_l \rangle} \langle v_l| \\
&= \underbrace{\sum_k |u_k\rangle\langle u_k|}_{=1} F \underbrace{\sum_l |v_l\rangle\langle v_l|}_{=1} \\
&= F. \qquad (1.2.39)
\end{aligned}$$

Indeed, we have succeeded in writing F as a function of U and V, with all

Us to the left of all Vs,

$$F = \sum_{k,l} |u_k\rangle\langle u_k| f(u_k, v_l) |v_l\rangle\langle v_l|$$

$$= \frac{1}{N^2} \sum_{k,l,m,n} (U/u_k)^m f(u_k, v_l)(V/v_l)^n$$

$$= \sum_{k,l} f_{kl} U^k V^l \tag{1.2.40}$$

with

$$f_{kl} = \frac{1}{N^2} \sum_{m,n} u_m^{-k} f(u_m, v_n) v_n^{-l} \tag{1.2.41}$$

after interchanging $k \leftrightarrow m$, $l \leftrightarrow n$ in the summation.

1-5 Show that

$$\mathrm{tr}\{U^k V^l\} = 0 \quad \text{unless} \quad k = l = 0 \pmod{N}$$

and then relate f_{kl} to $\mathrm{tr}\{U^{-k} F V^{-l}\}$.

1-6 For $F = \sum_{k,l} f_{kl} U^k V^l$ and $G = \sum_{k,l} g_{kl} U^k V^l$, express $\mathrm{tr}\{F^\dagger G\}$ in terms of the coefficients f_{kl} and g_{kl}.

With the general version of the Kronecker δ symbol,

$$\delta(x, y) = \begin{cases} 1 & \text{if} \quad x = y, \\ 0 & \text{if} \quad x \neq y, \end{cases} \tag{1.2.42}$$

we can write

$$|u_k\rangle\langle u_k| = \sum_j |u_j\rangle \delta(u_j, u_k)\langle u_j|$$

$$= \delta(U, u_k) \tag{1.2.43}$$

where the last step is an application of the general form of an operator function $f(A)$, the spectral decomposition in (1.1.17). Likewise we have

$$|v_l\rangle\langle v_l| = \delta(V, v_l), \tag{1.2.44}$$

and then

$$F = \sum_{k,l} \delta(U, u_k) f(u_k, v_l) \delta(V, v_l). \tag{1.2.45}$$

In view of the δ symbols, we can evaluate the sums over k and l and so arrive at

$$F = f(U; V) \tag{1.2.46}$$

where the semicolon is a reminder that all Us stand to the left of all Vs in this expression.

We return to (1.2.38) and reveal the following:

$$
\begin{aligned}
f(u_k, v_l) &= \frac{\langle u_k | F | v_l \rangle \langle v_l | u_k \rangle}{\langle u_k | v_l \rangle \langle v_l | u_k \rangle} \\
&= N \langle u_k | F | v_l \rangle \langle v_l | u_k \rangle \\
&= N \operatorname{tr} \{ | u_k \rangle \langle u_k | F | v_l \rangle \langle v_l | \} \\
&= N \operatorname{tr} \{ \delta(U, u_k) \, F \, \delta(V, v_l) \} \ ,
\end{aligned}
\tag{1.2.47}
$$

where we recall the defining property of the *trace*, that is

$$\operatorname{tr} \{ | 1 \rangle \langle 2 | \} = \langle 2 | 1 \rangle \tag{1.2.48}$$

for any ket $| 1 \rangle$ and any bra $\langle 2 |$. Relation (1.2.47) is the reciprocal to (1.2.45) inasmuch as we go from $F(U, V)$ to $f(u, v)$ in (1.2.47), and from $f(u, v)$ to $F(U, V)$ in (1.2.45).

We thus have a simple procedure for finding the U, V-ordered version of a given operator F:

First evaluate $f(u_k, v_l) = \dfrac{\langle u_k | F | v_l \rangle}{\langle u_k | v_l \rangle}$; then
replace $u_k \to U, \ v_l \to V$ with due attention
to the ordering in products, all U operators
must stand to the left of all V operators. $\qquad$ (1.2.49)

Here is an elementary example. For $F = VU$ we have

$$
\begin{aligned}
f(u_k, v_l) &= \frac{\langle u_k | VU | v_l \rangle}{\langle u_k | v_l \rangle} = \frac{\langle u_{k+1} | v_{l+1} \rangle}{\langle u_k | v_l \rangle} \\
&= \frac{\mathrm{e}^{\mathrm{i} \frac{2\pi}{N}(k+1)(l+1)} / \sqrt{N}}{\mathrm{e}^{\mathrm{i} \frac{2\pi}{N} kl} / \sqrt{N}} = \mathrm{e}^{\mathrm{i} \frac{2\pi}{N}(k+l+1)}
\end{aligned}
\tag{1.2.50}
$$

or

$$f(u_k, v_l) = u_k v_l \, \mathrm{e}^{\mathrm{i} \frac{2\pi}{N}} \tag{1.2.51}$$

so that

$$VU = F = UV \, \mathrm{e}^{\mathrm{i}\frac{2\pi}{N}} , \qquad\qquad (1.2.52)$$

which we know already.

1-7 What is the U, V-ordered version of $|u_j\rangle\langle u_k|$, of $|v_j\rangle\langle v_k|$?

1-8 We have

$$|u_k\rangle\langle u_k| = \delta(U, u_k) = \frac{1}{N} \sum_l (U/u_k)^l .$$

Therefore,

$$\sum_k \frac{1}{N} \sum_l (U/u_k)^l = 1 .$$

Verify this by a direct evaluation of this sum, that is: first sum over k, then over l.

1-9 We have

$$F = \sum_{k,l} f_{kl} U^k V^l = f(U; V) .$$

Express the trace of F in terms of the N^2 coefficients f_{kl}, and also in terms of the N^2 numbers $f(u_k, v_l)$.

1-10 Show that

(1) if F commutes with U, that is $FU = UF$, then F is
    a function of U only, $F = f(U)$;
(2) if F commutes with V, that is $FV = VF$, then F is
    a function of V only, $F = f(V)$.

What is implied for an operator F that commutes with both U and V? The answer to this question is *Schur's lemma*, named after Issai Schur.

1-11 Consider an arbitrary operator X and define F by

$$F = \sum_{k,l=1}^N V^{-l} U^{-k} X U^k V^l ,$$

that is: F is the sum of all N^2 operators obtained from X by the basic unitary transformations that result from repeated applications of U and V. Show that F commutes with U and with V. Then conclude that

$$F = N \operatorname{tr}\{X\} .$$

1-12 For odd N, that is $N = 2M + 1$ with $M = 1, 2, 3, \ldots$, we define N^2 operators in accordance with

$$W_{00} = \frac{1}{N} \sum_{j=-M}^{M} \sum_{k=-M}^{M} U^j V^k \, e^{i\pi jk/N} \quad \text{and} \quad W_{lm} = V^m U^{-l} W_{00} U^l V^{-m}$$

for $-M \leq l, m \leq M$. Show that all W_{lm}s are hermitian, and evaluate $\text{tr}\{W_{lm}\}$ and $\text{tr}\{W_{lm} W_{l'm'}\}$.

1-13 Establish that an arbitrary operator F can be written as a weighted sum of the W_{lm}s,

$$F = \frac{1}{N} \sum_{l,m} f_{lm} W_{lm} \quad \text{with} \quad f_{lm} = \text{tr}\{F W_{lm}\} \, ,$$

and express $\text{tr}\{F\}$ in terms of the coefficients f_{lm}.

1.2.3 Bohr's principle. Technical formulation

In summary, we have established a clear technical formulation of Niels H. D. Bohr's *principle of complementarity*:

> (1) for each quantum degree of freedom there
> is a pair of complementary observables
> and (2) all observables are functions of this pair.

This wording of the principle is a minor extension of the insights gained by Hermann K. H. Weyl and Julian Schwinger in their seminal work on quantum kinematics. We comment on the phenomenological implication of the complementarity principle in Section 1.2.7 below.

1.2.4 Composite degrees of freedom

In the above reasoning, the dimension N of the degree of freedom can be any integer number, prime or composite. In the case of composite numbers, we can regard the quantum degree of freedom as being composite as well, namely composed of simpler systems. It is sufficient to illustrate this in the example of $N = 6 = 2 \times 3$.

Here we have the periodic unitary operators

$$U^6 = 1 \quad \text{and} \quad V^6 = 1 \tag{1.2.53}$$

of period 6, but as

$$\left(U^3\right)^2 = 1, \quad \left(U^2\right)^3 = 1 \tag{1.2.54}$$

show, there are also operators with periods 2 and 3. This suggests that we can regard the $N = 6$ degree of freedom as composed of a $N = 2$ one and a $N = 3$ one.

To make this point more explicitly, let us arrange the six basis states $|a_1\rangle, \ldots, |a_6\rangle (\equiv |v_1\rangle, \ldots, |v_6\rangle)$ in a two-dimensional array

$$
\begin{array}{ccc}
1 & 2 & 3 \\
\bullet & \bullet & \bullet \\[1em]
\bullet & \bullet & \bullet \\
4 & 5 & 6
\end{array}
$$

and then relabel them accordingly,

$$
\begin{array}{ccc}
11 & 12 & 13 \\
\bullet & \bullet & \bullet \\[1em]
\bullet & \bullet & \bullet \\
21 & 22 & 23
\end{array}
$$

Rather than the original cyclic permutation $1 \to 2 \to 3 \to 4 \to 5 \to 6 \to 1$, we now have cyclic permutations of the rows $11 \leftrightarrow 21$ & $12 \leftrightarrow 22$ & $13 \leftrightarrow 23$, and of the columns $11 \to 12 \to 13 \to 11$ & $21 \to 22 \to 23 \to 21$:

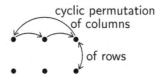

So, there are two operators effecting cyclic permutations, one with period 2 and the other with period 3,

$$U_1^2 = 1, \quad U_2^3 = 1, \tag{1.2.55}$$

which act on the respective indices,

$$
\begin{aligned}
U_1 |v_{jk}\rangle &= |v_{j+1\,k}\rangle, \\
U_2 |v_{jk}\rangle &= |v_{j\,k+1}\rangle,
\end{aligned}
\tag{1.2.56}
$$

where the first index j is modulo 2, the second index k is modulo 3. Clearly, U_1 commutes with U_2

$$U_1 U_2 = U_2 U_1 \,, \tag{1.2.57}$$

and it is easy to show (do this yourself) that this is also true for their complementary partners V_1 and V_2. That is,

$$V_1 V_2 = V_2 V_1 \,, \quad U_1 V_2 = V_2 U_1 \,, \quad V_1 U_2 = U_2 V_1 \,. \tag{1.2.58}$$

But the pairs U_1, V_1 and U_2, V_2 are pairs of complementary observables of the $N = 2$ and $N = 3$ types, respectively,

$$U_1 V_1 = \mathrm{e}^{-\mathrm{i}\frac{2\pi}{2}} V_1 U_1 \,, \quad U_2 V_2 = \mathrm{e}^{-\mathrm{i}\frac{2\pi}{3}} V_2 U_2 \,. \tag{1.2.59}$$

In short: U_1, V_1 and U_2, V_2 refer to independent degrees of freedom, the prime degrees of freedom that are the constituents of the composite degree of freedom with $N = 6$.

1-14   Show that $U = U_1 U_2$ has period 6. Begin with $|a_1\rangle \equiv |v_{11}\rangle$ and find $|a_2\rangle, \ldots, |a_6\rangle$ in accordance with (1.2.2).

In the general situation of a composite value of N other than 6, this process of factorization can be repeated until the given $N = N_1 N_2 \cdots N_n$ is broken up into the prime degrees of freedom of which it is composed, or can be thought of as being composed. As a consequence, the elementary quantum degrees of freedom have N values that are prime and cannot be decomposed further.

1.2.5   The limit $N \to \infty$. Symmetric case

The primes $N = 2, 3, 5, 7, 11, 13, \ldots$ are all odd, except for $N = 2$, so that we can restrict ourselves to odd N values in the limit $N \to \infty$. It is then possible to change from the numbering

$$k = 1, 2, \ldots, N \tag{1.2.60}$$

to a new numbering

$$k = 0, \pm 1, \pm 2, \ldots, \pm \frac{N-1}{2} \,. \tag{1.2.61}$$

Further, as N grows, the basic unit of complex phase $2\pi/N$ gets arbitrarily small. We introduce a small quantity, ϵ, to deal with this,

$$\frac{2\pi}{N} = \epsilon^2 \,. \tag{1.2.62}$$

Aiming at a continuous degree of freedom in the limit $N \to \infty$, we also relabel the states in the sense that

$$k \longrightarrow k\epsilon = x = 0, \pm\epsilon, \pm 2\epsilon, \dots, \pm\left(\frac{\pi}{\epsilon} - \frac{\epsilon}{2}\right),$$

$$l \longrightarrow l\epsilon = p = 0, \pm\epsilon, \pm 2\epsilon, \dots, \pm\left(\frac{\pi}{\epsilon} - \frac{\epsilon}{2}\right). \tag{1.2.63}$$

Then the numbers x and p will cover the real axis, $-\infty < x, p < \infty$, when $N \to \infty$, $\epsilon \to 0$.

The unitary operator U acting on $\left| v_l \right\rangle$ increases l by 1, so that it effects $p \to p + \epsilon$. Likewise V applied to $\left\langle u_k \right|$ results in $x \to x + \epsilon$. This suggests the identification of hermitian operators X and P such that

$$U = e^{i\epsilon X} \quad \text{with} \quad X = X^\dagger \,,$$

$$V = e^{i\epsilon P} \quad \text{with} \quad P = P^\dagger \,. \tag{1.2.64}$$

The Weyl commutator (1.2.36),

$$U^l V^k = e^{-i\frac{2\pi}{N}kl} V^k U^l \,, \tag{1.2.65}$$

then appears as

$$e^{il\epsilon X} e^{ik\epsilon P} = e^{-ik\epsilon l\epsilon} e^{ik\epsilon P} e^{il\epsilon X} \,, \tag{1.2.66}$$

that is

$$e^{ipX} e^{ixP} = e^{-ixp} e^{ixP} e^{ipX} \,. \tag{1.2.67}$$

The two equivalent versions

$$e^{-ixP} e^{ipX} e^{ixP} = e^{ip(X - x)} \,,$$

$$e^{ipX} e^{ixP} e^{-ipX} = e^{ix(P - p)} \tag{1.2.68}$$

look much more conspicuous after we use the identity

$$U^{-1} f(A) U = f(U^{-1} A U) \,, \tag{1.2.69}$$

which — as we recall — is valid for any operator function $f(A)$ and any unitary operator U, twice to establish

$$
\begin{aligned}
e^{ip(e^{-ixP}X\,e^{ixP})} &= e^{ip(X-x)}\,, \\
e^{ix(e^{ipX}P\,e^{-ipX})} &= e^{ix(P-p)}\,.
\end{aligned}
\qquad (1.2.70)
$$

It is tempting to conclude that

$$
\begin{aligned}
e^{-ixP}X\,e^{ixP} &= X - x\,, \\
e^{ipX}P\,e^{-ipX} &= P - p\,,
\end{aligned}
\qquad (1.2.71)
$$

but this does not follow without imposing a restricting condition, just as

$$
e^{i\alpha} = e^{i\beta} \quad (\alpha, \beta:\ \text{two real numbers}) \qquad (1.2.72)
$$

does not imply $\alpha = \beta$, but only that $\alpha - \beta$ is an integer multiple of 2π.

To understand the restricting condition, we visualize the cyclic permutation by a circle:

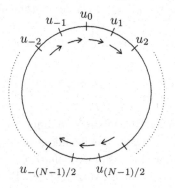

where an application of V turns the wheel of u_k states by one notch. For large N, small ϵ, the picture is this

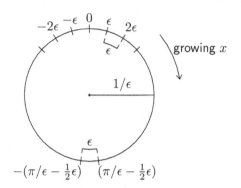

The circumference of the circle is $N\epsilon = 2\pi/\epsilon$, so that the radius is $1/\epsilon$ and becomes infinitely large as $N \to \infty$, $\epsilon \to 0$. In this limit then, any *finite* portion of the circle is indistinguishable from a straight line,

If we thus restrict ourselves to situations in which only a finite range of x values matters, thereby explicitly giving up the periodicity that would force us to identify $x = +\infty$ with $x = -\infty$, the statements in (1.2.71) become correct.

By comparing terms that are of first order in x or p in (1.2.71) we get

$$XP - PX = i \quad \text{or} \quad [X, P] = i, \tag{1.2.73}$$

which is, of course, Werner Heisenberg's commutation relation for position X and momentum P, here for dimensionless operators, rather than the normal ones with metrical dimensions of length and mass × velocity. As a consequence, Planck's constant $\hbar$ (Max K. E. L. Planck) does not appear on the right-hand side.

With this commutator established by the 1st-order terms, all higher-order terms take care of themselves, which is seen most easily by differentiation. Consider, say, the first statement in (1.2.71), where we get

$$\frac{\partial}{\partial x}\left(e^{-ixP}X\,e^{ixP}\right) = e^{-ixP}\underbrace{(-iPX + XiP)}_{=\,i[X,P]\,=\,-1}e^{ixP}$$

$$= -1 \qquad (1.2.74)$$

on the left and

$$\frac{\partial}{\partial x}(X - x) = -1 \qquad (1.2.75)$$

on the right. Thus, these are two functions of x, which have the same derivative everywhere and agree in the vicinity of $x = 0$. It follows that the functions are the same for all values of x.

These $N \to \infty$ considerations for U and V have a counterpart in their respective kets and bras. It should be reasonably clear, given what we know about the X, P pair from *Basic Matters* and *Simple Systems*, or any other introductory text, that

$$\langle u_k | v_l \rangle = \frac{1}{\sqrt{N}}\,e^{i\frac{2\pi}{N}kl} \qquad (1.2.76)$$

turns into

$$\langle x | p \rangle = \frac{1}{\sqrt{2\pi}}\,e^{ixp}, \qquad (1.2.77)$$

and the orthonormality and completeness relations

$$\langle u_k | u_{k'} \rangle = \delta_{kk'}, \qquad \sum_k |u_k\rangle\langle u_k| = 1 \qquad (1.2.78)$$

become

$$\langle x | x' \rangle = \delta(x - x'), \qquad \int dx\,|x\rangle\langle x| = 1, \qquad (1.2.79)$$

and likewise for the momentum states, with consistent replacements of Kronecker δ symbols by Dirac δ functions (Paul A. M. Dirac), and summations by integrations.

More specifically, we need to identify

$$\langle x| = \frac{1}{\sqrt{\epsilon}}\langle u_k|\Big|_{\epsilon\,\to\,0} \qquad \text{with} \quad k\epsilon = x \qquad (1.2.80)$$

and

$$|p\rangle = |v_l\rangle \frac{1}{\sqrt{\epsilon}}\bigg|_{\epsilon \to 0} \qquad \text{with} \quad l\epsilon = p, \qquad (1.2.81)$$

and then we have

$$\frac{1}{\epsilon}\langle u_k|v_l\rangle \xrightarrow[\epsilon \to 0]{} \langle x|p\rangle. \qquad (1.2.82)$$

As one example of many that could be used for illustration equally well, let us take a look at the transition from

$$\delta(U, u_k) = \frac{1}{N}\sum_{l(\neq k)}\left(U/u_k\right)^l = |u_k\rangle\langle u_k| \qquad (1.2.83)$$

to

$$\delta(X - x) = \int \frac{\mathrm{d}p}{2\pi}\, \mathrm{e}^{\mathrm{i}p(X - x)} = |x\rangle\langle x|. \qquad (1.2.84)$$

We proceed from

$$\begin{aligned}
\frac{1}{\epsilon}\delta(U, u_k) &= \frac{1}{N\epsilon}\sum_{l(\neq k)}\left(\mathrm{e}^{\mathrm{i}\epsilon X}\, \mathrm{e}^{-\mathrm{i}\epsilon x}\right)^l \\
&= \frac{1}{N\epsilon^2}\sum_{l(\neq k)}\underbrace{\Delta p}\, \mathrm{e}^{\mathrm{i}p(X - x)} \\
&\quad\; {\scriptstyle = 1/(2\pi)}
\end{aligned} \qquad (1.2.85)$$

with $p = l\epsilon$ and $\Delta p = \epsilon$ for the difference between successive p values. Upon recognizing that the l summation is the Riemann sum (G. F. Bernhard Riemann) for the integral in

$$\frac{1}{\epsilon}\delta(U, u_k) = \frac{1}{2\pi}\int_{\epsilon/2 - \pi/\epsilon}^{\pi/\epsilon - \epsilon/2} \mathrm{d}p\, \mathrm{e}^{\mathrm{i}p(X - x)}, \qquad (1.2.86)$$

the limit $\epsilon \to 0$ is immediate,

$$\frac{1}{\epsilon}\delta(U, u_k) \xrightarrow[\epsilon \to 0]{} \frac{1}{2\pi}\int_{-\infty}^{\infty} \mathrm{d}p\, \mathrm{e}^{\mathrm{i}p(X - x)} = \delta(X - x). \qquad (1.2.87)$$

The familiar Fourier representation (Jean B. J. Fourier) of the Dirac δ function establishes the last identity. The other limit offered by (1.2.83) in conjunction with (1.2.80),

$$\frac{1}{\epsilon}\delta(U, u_k) = \left(|u_k\rangle\frac{1}{\sqrt{\epsilon}}\right)\left(\frac{1}{\sqrt{\epsilon}}\langle u_k|\right) \xrightarrow[\epsilon \to 0]{} |x\rangle\langle x|, \qquad (1.2.88)$$

is consistent with what we get from the spectral decomposition of $\delta(X - x)$,

$$\delta(X - x) = \int dx' \, |x'\rangle\delta(x' - x)\langle x'| = |x\rangle\langle x|, \qquad (1.2.89)$$

as it should be.

It is natural and convenient to use dimensionless operators X and P for this study of the limit $N \to \infty$, but eventually we want to have the correct metrical dimensions of length for position X and mass $\times$ velocity for momentum P. All that is needed is the introduction of Planck's constant $\hbar$ in the right places, such as

$$[X, P] = i\hbar \qquad (1.2.90)$$

for the Heisenberg commutator rather than (1.2.73), and

$$\langle x|p\rangle = \frac{e^{ixp/\hbar}}{\sqrt{2\pi\hbar}} \qquad (1.2.91)$$

for the xp transformation function rather than (1.2.77). The orthonormality and completeness relations for the position states in (1.2.79) continue to hold without change, but we should take note of the metrical dimension $(\text{length})^{-\frac{1}{2}}$ of $|x\rangle$ and $\langle x|$. Corresponding remarks apply to the momentum states.

For the record and future reference it is worth recalling the basic relations between commutators and differentiations,

$$[X, f(X, P)] = i\hbar\frac{\partial f(X, P)}{\partial P},$$
$$[f(X, P), P] = i\hbar\frac{\partial f(X, P)}{\partial X}, \qquad (1.2.92)$$

where $f(X, P)$ is any well defined function of X and P. These generalizations of the Heisenberg commutator are in fact implications of (1.2.90), and in turn contain (1.2.90) as special cases.

1-15 Combine (1.2.71) with (1.2.69) to first establish that

$$e^{-ixP/\hbar} f(X,P) e^{ixP/\hbar} = f(X-x,P)$$
$$\text{and} \quad e^{ipX/\hbar} f(X,P) e^{-ipX/\hbar} = f(X,P-p),$$

and then use this in

$$\frac{\partial f(X,P)}{\partial X} = \frac{1}{x}\left[f(X,P) - f(X-x,P)\right]\Bigg|_{x \to 0},$$

for example, to derive (1.2.92).

1.2.6   The limit $N \to \infty$. Asymmetric case

The limit $N \to \infty$ discussed in Section 1.2.5 is *symmetric* inasmuch as U and V are treated on completely equal footing. This symmetric procedure is, however, not the only way of performing the limit $N \to \infty$. There are, in fact, three important *asymmetric* limits, of which we shall consider one.

This time we start with the relation

$$\delta_{kl} = \sum_m \langle u_k | v_m \rangle \langle v_m | u_l \rangle \tag{1.2.93}$$

which states the completeness of the V states and the orthonormality of the U states. In

$$\langle u_k | v_m \rangle = \frac{1}{\sqrt{N}} e^{i\frac{2\pi}{N}km} = \frac{1}{\sqrt{N}} \left(e^{i\frac{2\pi}{N}k} \right)^m \tag{1.2.94}$$

we encounter a phase factor

$$e^{i\frac{2\pi}{N}k} = e^{i\phi} \quad \text{with} \quad \phi = \frac{2\pi}{N}k \quad \text{and} \quad k = 0,\ldots,N-1. \tag{1.2.95}$$

The phases $\phi = \dfrac{2\pi}{N}k$ and $\phi' = \dfrac{2\pi}{N}l$ will cover the whole range

$$0 \leq \phi, \phi' < 2\pi \tag{1.2.96}$$

densely in the limit $N \to \infty$. We note immediately that the relation

$$e^{i\frac{2\pi}{N}k} = e^{i\phi} \tag{1.2.97}$$

identifies ϕ only up to an arbitrary multiple of 2π, so that there is no point in distinguishing between ϕ and $\phi + 2\pi$.

For the m summation we choose $m = 0, \pm 1, \ldots, \pm\frac{1}{2}(N-1)$, so that

$$
N\delta_{kl} = \sum_{m=-\frac{1}{2}(N-1)}^{\frac{1}{2}(N-1)} e^{im(\phi - \phi')} = \left\{ \begin{array}{ll} N & \text{if} \quad \phi = \phi' \pmod{2\pi} \\ 0 & \text{if} \quad \phi \neq \phi' \pmod{2\pi} \end{array} \right\}
$$

$$
= N\delta(\phi, \phi') \pmod{2\pi}
$$

$$
= N \sum_{n=-\infty}^{\infty} \delta(\phi, \phi' + 2\pi n). \qquad (1.2.98)
$$

In the limit $N \to \infty$, the sum is

$$
\sum_{m=-\infty}^{\infty} e^{im(\phi - \phi')} = 2\pi \sum_{n=-\infty}^{\infty} \delta(\phi - \phi' - 2\pi n). \qquad (1.2.99)
$$

This so-called *Poisson identity*, which is named after Siméon-Denis Poisson, can be verified by comparing the Fourier coefficients of these two periodic functions of ϕ. On the left we get

$$
\int_0^{2\pi} \frac{d\phi}{2\pi} e^{-ij\phi} \sum_{m=-\infty}^{\infty} e^{im(\phi - \phi')} = e^{-ij\phi'}, \qquad (1.2.100)
$$

and on the right

$$
\int_0^{2\pi} \frac{d\phi}{2\pi} e^{-ij\phi} 2\pi \sum_{n=-\infty}^{\infty} \delta(\phi - \phi' - 2\pi n) = e^{-ij\phi'}. \qquad (1.2.101)
$$

Indeed, all Fourier coefficients are identical and, therefore, the functions are the same.

We are thus invited to introduce bras $\langle \phi|$ and kets $|m\rangle$ in accordance with

$$
\langle \phi| = \langle \phi + 2\pi| \quad \text{(periodic)} \qquad (1.2.102)
$$

and

$$
\langle \phi|m\rangle = \frac{1}{\sqrt{2\pi}} e^{im\phi}. \qquad (1.2.103)
$$

These are such that

$$
\langle \phi|\phi'\rangle = \sum_{m=-\infty}^{\infty} \langle \phi|m\rangle\langle m|\phi'\rangle
$$

$$
= \frac{1}{2\pi} \sum_m e^{im(\phi - \phi')} = \sum_n \delta(\phi - \phi' - 2\pi n) \qquad (1.2.104)
$$

and

$$\langle m|m'\rangle = \int_{(2\pi)} \mathrm{d}\phi\, \langle m|\phi\rangle\langle\phi|m'\rangle$$

$$= \int_{(2\pi)} \frac{\mathrm{d}\phi}{2\pi}\, \mathrm{e}^{-\mathrm{i}(m-m')\phi} = \delta_{mm'} \tag{1.2.105}$$

state the orthonormality and completeness of both sets of vectors. The ϕ integration in (1.2.105) covers any interval of 2π, such as $0\cdots 2\pi$ or $-\pi\cdots\pi$.

In the unitary operator

$$\sum_{m=-\infty}^{\infty} |m\rangle\, \mathrm{e}^{\mathrm{i}m\varphi}\langle m| = \mathrm{e}^{\mathrm{i}\varphi L/\hbar} \tag{1.2.106}$$

we recognize the hermitian operator

$$L = \sum_{m=-\infty}^{\infty} |m\rangle\hbar m\langle m| \tag{1.2.107}$$

that generates shifts of ϕ,

$$\langle\phi| \longrightarrow \langle\phi|\, \mathrm{e}^{\mathrm{i}\varphi L/\hbar} = \sum_m \langle\phi|m\rangle\, \mathrm{e}^{\mathrm{i}m\varphi}\langle m|$$

$$= \sum_m \underbrace{\frac{1}{\sqrt{2\pi}}\, \mathrm{e}^{\mathrm{i}m\phi}\, \mathrm{e}^{\mathrm{i}m\varphi}}_{= \mathrm{e}^{\mathrm{i}m(\phi+\varphi)}/\sqrt{2\pi} = \langle\phi+\varphi|m\rangle}\langle m|$$

$$= \sum_m \langle\phi+\varphi|m\rangle\langle m| = \langle\phi+\varphi|\,, \tag{1.2.108}$$

which are, geometrically speaking, rotations around a fixed axis:

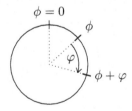

As it should be, this picture is consistent with the periodic nature of the bras $\langle\phi|$. We conclude that L is the angular momentum operator associated with the rotation around this axis.

1-16 Consider the unitary operators

$$U^n = \int_{(2\pi)} d\phi \, |\phi\rangle \, e^{in\phi} \langle\phi| = \left(\int_{(2\pi)} d\phi \, |\phi\rangle \, e^{i\phi} \langle\phi| \right)^n$$

where, as in (1.2.105), the integration is over any 2π interval and n is any integer, positive or negative. What is $U^n |m\rangle$?

1-17 Compare $U^n e^{i\varphi L/\hbar}$ with $e^{i\varphi L/\hbar} U^n$. Do you get what you antici-pate?

The differential statement corresponding to

$$\langle \phi + \varphi | = \langle \phi | \, e^{i\varphi L/\hbar} \qquad (1.2.109)$$

is

$$\frac{\partial}{\partial \varphi} \langle \phi + \varphi | \bigg|_{\varphi = 0} = \langle \phi | \frac{i}{\hbar} L \qquad (1.2.110)$$

or

$$\frac{\hbar}{i} \frac{\partial}{\partial \phi} \langle \phi | = \langle \phi | L . \qquad (1.2.111)$$

This differential-operator representation of the angular momentum operator acting on an azimuth bra is the analog of

$$\frac{\hbar}{i} \frac{\partial}{\partial x} \langle x | = \langle x | P , \qquad (1.2.112)$$

Erwin Schrödinger's relation for position bras and the operator of linear momentum. Indeed, in the context of orbital angular momentum the cor-respondence $L \to \frac{\hbar}{i} \frac{\partial}{\partial \phi}$ appears in Section 4.3 of *Simple Systems*.

1-18 Get (1.2.112) as an implication of (1.2.91).

In closing this subject, let us note that, in addition to the X, P-type of continuous degree of freedom and the ϕ, L-type, there is also one for radial motion (position limited to positive values, corresponding momentum takes on all real values), and one for the polar angle or confinement to a finite range (position limited to a finite range, but without periodic values as for the ϕ variable, and momentum takes on all real values). All of them can be obtained as suitable limits $N \to \infty$.

1.2.7   Bohr's principle. Quantum indeterminism

Pairs of complementary observables are such that if the value of one observable is known, all outcomes are equally probable in a measurement of the second observable. Put differently, if one observable is sharp, the other is completely undetermined. The situation is clearly an extreme case of *quantum indeterminism*. But, as extreme as it may seem, it is not at all untypical. In fact, it is the generic situation, because there are always undetermined observables, irrespective of the preparation of the system.

Consider, therefore, the most general case, that is a quantum system prepared in a state described by a statistical operator ρ; see Section 2.15 in *Basic Matters* and Section 1.6 in *Simple Systems*. We write ρ in its diagonal form,

$$\rho = \sum_{j=1}^{N} |r_j\rangle r_j \langle r_j|$$

with $\quad r_j \geq 0, \quad \sum_j r_j = 1, \quad \langle r_j|r_k\rangle = \delta_{jk}.$ \qquad (1.2.113)

Then define

$$|a_k\rangle = \frac{1}{\sqrt{N}} \sum_{j=1}^{N} |r_j\rangle \, e^{i\frac{2\pi}{N}jk} \qquad (1.2.114)$$

so that

$$\langle a_k|a_l\rangle = \delta_{kl} \qquad (1.2.115)$$

as we know from (1.1.27). The observable introduced by

$$A = \sum_{k=1}^{N} |a_k\rangle k \langle a_k| \qquad (1.2.116)$$

has values $k = 1, 2, \ldots, N$, and all of them have probability $1/N$, because

$$\langle a_k|\rho|a_k\rangle = \sum_{j=1}^{N} r_j \underbrace{\left|\langle a_k|r_j\rangle\right|^2}_{=1/N} = \frac{1}{N} \sum_{j=1}^{N} r_j = \frac{1}{N}. \qquad (1.2.117)$$

Since ρ was quite arbitrary, and this construction for A is always possible, the above assertion is correct indeed, that is:

> Irrespective of the preparation of the quantum
> system, there are always observables that are
> completely undetermined.

One can regard this statement as another way of phrasing Bohr's principle of complementarity, a phrasing that emphasizes the phenomenology.

1-19 Guided by the φ, L example of Section 1.2.6, construct an undetermined observable for the case of $N = \infty$.

1.3 Brief review of basic dynamics

1.3.1 *Equations of motion*

Up to here, we have been reviewing quantum kinematics, that is the description of quantum systems. Now we turn to quantum dynamics, that is the evolution in time of quantum systems. One basic equation is Erwin Schrödinger's equation of motion, the *Schrödinger equation*, which we state both for bras and for kets,

$$i\hbar \frac{\partial}{\partial t}\langle \ldots, t| = \langle \ldots, t| H\big(A(t), t\big),$$

$$-i\hbar \frac{\partial}{\partial t}|\ldots, t\rangle = H\big(A(t), t\big)|\ldots, t\rangle. \tag{1.3.1}$$

The ellipses indicate fixed quantum numbers that identify the kets and bras, and $H\big(A(t), t\big)$ is the hermitian Hamilton operator at time t, regarded as a function of the dynamical variables $A(t)$, pairs of complementary observables for the various degrees of freedom under consideration, and perhaps of t itself.

All other operators are of the same general form, $F = F\big(A(t), t\big)$, and obey Werner Heisenberg's equation of motion, the *Heisenberg equation*,

$$\frac{\mathrm{d}}{\mathrm{d}t}F = \frac{\partial F}{\partial t} + \frac{1}{i\hbar}[F, H] \tag{1.3.2}$$

where $\dfrac{\mathrm{d}}{\mathrm{d}t}$ differentiates t globally

$$\frac{\mathrm{d}}{\mathrm{d}t}F\big(A(t), t\big), \tag{1.3.3}$$

whereas $\dfrac{\partial}{\partial t}$ means the *parametric* time derivative only,

$$\frac{\partial}{\partial t} F\big(A(t),t\big) .\qquad\qquad (1.3.4)$$

$$\underset{\text{only}}{\underbrace{\qquad\qquad}}{\uparrow}$$

Their difference

$$\left(\frac{\mathrm{d}}{\mathrm{d}t} - \frac{\partial}{\partial t}\right) F\big(A(t),t\big) = \frac{1}{\mathrm{i}\hbar}\big[F\big(A(t),t\big), H\big(A(t),t\big)\big] \qquad (1.3.5)$$

is the *dynamical* time derivative that originates in the dynamics of the system, that is the physical interactions of the constituents. Perhaps consult Chapter 3 of *Basic Matters* and Chapter 2 of *Simple Systems*, if you are uncertain about these matters.

We recall that there are some special cases. First, the Hamilton operator itself has no dynamical time dependence,

$$\frac{\mathrm{d}}{\mathrm{d}t} H = \frac{\partial}{\partial t} H , \qquad\qquad (1.3.6)$$

so that it is constant in time if there is no parametric time dependence,

$$H\big(A(t)\big) = H\big(A(t_0)\big) \quad \text{if} \quad \frac{\partial H}{\partial t} = 0 . \qquad (1.3.7)$$

Second, the dynamical variables themselves have only a dynamical time dependence,

$$\frac{\partial}{\partial t} A = 0 , \quad \frac{\mathrm{d}}{\mathrm{d}t} A = \frac{1}{\mathrm{i}\hbar}[A, H] . \qquad (1.3.8)$$

In particular, for position X and momentum P as the dynamical variables, we have, by virtue of (1.2.92),

$$\frac{\mathrm{d}}{\mathrm{d}t} X = \frac{\partial H}{\partial P} \quad \text{and} \quad \frac{\mathrm{d}}{\mathrm{d}t} P = -\frac{\partial H}{\partial X} , \qquad (1.3.9)$$

which have the same form as William R. Hamilton's equations of motion in classical mechanics.

Third, the statistical operator $\rho(A(t), t)$ has no total time dependence,

$$\frac{\mathrm{d}}{\mathrm{d}t} \rho = 0 , \quad \rho\big(A(t),t\big) = \rho\big(A(t_0), t_0\big) . \qquad (1.3.10)$$

This is to say that the parametric t dependence of ρ compensates fully for the dynamical t dependence. Therefore, the statistical operator obeys

$$\frac{\partial}{\partial t}\rho = -\frac{1}{\mathrm{i}\hbar}[\rho, H] ; \qquad (1.3.11)$$

this special case of the Heisenberg equation (1.3.2) is the so-called *von Neumann equation*, named after John von Neumann. It is the quantum analog of Joseph Liouville's equation of motion in classical statistical physics.

1.3.2 Time transformation functions

The descriptions at different times are related to each other by the time transformation functions, such as $\langle a, t_1 | b, t_2 \rangle$ for kets $|b\rangle$ at the early time t_2 and bras $\langle a|$ at the late time t_1. For example, if the state of the system is specified by the probability amplitudes $\langle b, t_2 | \ \rangle$ at the early time t_2,

$$| \ \rangle = \sum_b |b, t_2\rangle\langle b, t_2 | \ \rangle, \qquad (1.3.12)$$

we find the amplitudes $\langle a, t_1 | \ \rangle$ by means of

$$\langle a, t_1 | \ \rangle = \sum_b \langle a, t_1 | b, t_2 \rangle\langle b, t_2 | \ \rangle, \qquad (1.3.13)$$

which follows from applying both sides of (1.3.12) to $\langle a, t_1|$. We can examine the evolution of any given state as soon as we know the time transformation functions.

Being conscious of the fact that all fundamental evolution equations in physics are differential equations, let us ask the following question:

> How does $\langle a, t_1 | b, t_2 \rangle$ change if there is a small change
> in the Hamilton operator at the intermediate time t?

The primary effect of such a change in H at time t is on $\langle a', t + \mathrm{d}t | b', t \rangle$, namely

$$\delta\langle a', t + \mathrm{d}t | b', t \rangle = \delta\left[\langle a', t | \left(1 - \frac{\mathrm{i}}{\hbar}H(t)\mathrm{d}t \right) |b', t\rangle \right]$$

$$= \langle a', t | \left(-\frac{\mathrm{i}}{\hbar}\delta H(t)\mathrm{d}t \right) |b', t\rangle$$

$$= \langle a', t + \mathrm{d}t | \left(-\frac{\mathrm{i}}{\hbar}\delta H(t)\mathrm{d}t \right) |b', t\rangle \qquad (1.3.14)$$

where $H(t) \equiv H\big(A(t), t\big)$ for brevity and the last step recognizes that we are only dealing with terms that are of first order in the time increment dt. Thus, the effect on

$$\langle a, t_1 | b, t_2 \rangle = \sum_{a', b'} \langle a, t_1 | a', t + dt \rangle \langle a', t + dt | b', t \rangle \langle b', t | b, t_2 \rangle \qquad (1.3.15)$$

is

$$\delta \langle a, t_1 | b, t_2 \rangle$$
$$= \sum_{a', b'} \langle a, t_1 | a', t + dt \rangle \langle a', t + dt | \left(-\frac{i}{\hbar} \delta H(t) dt \right) | b', t \rangle \langle b', t | b, t_2 \rangle$$
$$= \langle a, t_1 | \left(-\frac{i}{\hbar} \delta H(t) dt \right) | b, t_2 \rangle . \qquad (1.3.16)$$

This is the contribution of an infinitesimal change of H, an infinitesimal change of the dynamics, at the particular intermediate time t, and we get the accumulated effect of small changes at all intermediate times by integration,

$$\delta \langle a, t_1 | b, t_2 \rangle = \langle a, t_1 | \int_{t_2}^{t_1} \left(-\frac{i}{\hbar} \delta H(t) dt \right) | b, t_2 \rangle . \qquad (1.3.17)$$

As an elementary example, let us consider the mass dependence of the time transformation function

$$\langle x, t_1 | x', t_2 \rangle = \sqrt{\frac{M}{i 2 \pi \hbar T}} \, e^{\frac{i}{\hbar} \frac{M}{2T} (x - x')^2} , \qquad (1.3.18)$$

where $T = t_1 - t_2$ is the total duration and the underlying Hamilton operator

$$H = \frac{1}{2M} P^2 \qquad (1.3.19)$$

is that of force-free motion in one dimension. We have

$$\delta_M H(t) = -\frac{\delta M}{2M^2} P(t)^2 \qquad (1.3.20)$$

where

$$P(t) = \frac{M}{T} \Big(X(t_1) - X(t_2) \Big) = P(t_1) = P(t_2) \qquad (1.3.21)$$

is constant in time (no force — no change of the momentum).

We wish to write

$$\delta_M H = -\frac{\delta M}{2M^2}\left(\frac{M}{T}\right)^2\left(X(t_1) - X(t_2)\right)^2 \tag{1.3.22}$$

$$= -\frac{\delta M}{2T^2}\left(X(t_1)^2 + X(t_2)^2 - X(t_1)X(t_2) - X(t_2)X(t_1)\right)$$

with the position operators $X(t_1), X(t_2)$ in their natural order: $X(t_1)$ to the left, $X(t_2)$ to the right, so that they will stand next to their respective eigenstates, namely bra $\langle x, t_1|$ for $X(t_1)$ and ket $|x', t_2\rangle$ for $X(t_2)$. We thus need the commutator

$$[X(t_1), X(t_2)] = \left[X(t_2) + \frac{T}{M}P(t_2), X(t_2)\right]$$

$$= -i\hbar\frac{T}{M} \tag{1.3.23}$$

in

$$X(t_2)X(t_1) = X(t_1)X(t_2) - [X(t_1), X(t_2)]$$

$$= X(t_1)X(t_2) + i\hbar\frac{T}{M}. \tag{1.3.24}$$

Accordingly,

$$\delta_M H = -\frac{\delta M}{2T^2}\left(X(t_1)^2 + X(t_2)^2 - 2X(t_1)X(t_2) - i\hbar\frac{T}{M}\right) \tag{1.3.25}$$

and

$$\delta_M\langle x, t_1|x', t_2\rangle = \langle x, t_1|x', t_2\rangle \int_{t_2}^{t_1} dt\left(-\frac{i}{\hbar}\right)\left(-\frac{\delta M}{2T^2}\right)\left[(x - x')^2 - i\hbar\frac{T}{M}\right]$$

$$= \langle x, t_1|x', t_2\rangle\left[\frac{i}{\hbar}\frac{\delta M}{2T}(x - x')^2 + \frac{\delta M}{2M}\right], \tag{1.3.26}$$

implying first

$$\delta_M\log\langle x, t_1|x', t_2\rangle = \frac{\delta_M\langle x, t_1|x', t_2\rangle}{\langle x, t_1|x', t_2\rangle}$$

$$= \frac{i}{\hbar}\frac{\delta M}{2T}(x - x')^2 + \frac{\delta M}{2M}$$

$$= \delta_M\left(\frac{i}{\hbar}\frac{M}{2T}(x - x')^2 + \log\sqrt{M}\right) \tag{1.3.27}$$

and then

$$\langle x, t_1 | x', t_2 \rangle \propto \sqrt{M} \, e^{\frac{i}{\hbar} \frac{M}{2T} (x - x')^2} \qquad (1.3.28)$$

where the proportionality factor does not depend on M. We compare this with the known form of $\langle x, t_1 | x', t_2 \rangle$ in (1.3.18) and confirm that the M dependence thus found is correct.

1.4 Schwinger's quantum action principle

In view of (1.2.92) and (1.3.1), we also know how to deal with changes of the initial and final values of x and t,

$$\delta\langle x, t_1 | = \langle x, t_1 | \frac{i}{\hbar} \Big(P(t_1)\delta x - H(t_1)\delta t_1 \Big),$$

$$\delta | x', t_2 \rangle = -\frac{i}{\hbar} \Big(P(t_2)\delta x' - H(t_2)\delta t_2 \Big) | x', t_2 \rangle, \qquad (1.4.1)$$

which we abbreviate as

$$\delta\langle x, t_1 | = \langle x, t_1 | \frac{i}{\hbar} G_1 \qquad (1.4.2)$$

and

$$\delta | x', t_2 \rangle = -\frac{i}{\hbar} G_2 | x', t_2 \rangle \qquad (1.4.3)$$

where G_1, G_2 are the appropriate *generators* for infinitesimal changes of the bras and kets. Their specifc form depends on the quantum numbers that characterize the initial and final states. For example, in the case of an initial momentum ket, we have

$$\delta | p, t_2 \rangle = \Big(\frac{i}{\hbar} X(t_2)\delta p + \frac{i}{\hbar} H(t_2)\delta t_2 \Big) | p, t_2 \rangle$$

$$= -\frac{i}{\hbar} G_2 | p, t_2 \rangle \qquad (1.4.4)$$

with

$$G_2 = -X(t_2)\delta p - H(t_2)\delta t_2 \,. \qquad (1.4.5)$$

Quite generally, then, the response of a time transformation function to variations of both the initial and final states and the dynamics at intermediate times is

$$\delta\langle a, t_1 | b, t_2 \rangle = \frac{i}{\hbar} \langle a, t_1 | \left(G_1 - G_2 - \int_{t_2}^{t_1} dt\, \delta H(t) \right) | b, t_2 \rangle. \qquad (1.4.6)$$

Upon recognizing that we can derive the infinitesimal operators as variations of an *action* W_{12},

$$G_1 - G_2 - \int_{t_2}^{t_1} dt\, \delta H(t) = \delta W_{12}, \qquad (1.4.7)$$

this becomes Julian Schwinger's *quantum action principle*

$$\delta\langle a, t_1 | b, t_2 \rangle = \frac{i}{\hbar} \langle a, t_1 | \delta W_{12} | b, t_2 \rangle. \qquad (1.4.8)$$

The particular form of W_{12} depends thereby on the form of the generators G_1 and G_2 that are needed for the specified bras and kets. In particular, we have

$$W_{12} = \int_{t_2}^{t_1} dt \left(P(t) \frac{dX}{dt} - H(t) \right) \qquad (1.4.9)$$

for

$$\delta\langle x, t_1 | x', t_2 \rangle = \frac{i}{\hbar} \langle x, t_1 | \delta W_{12} | x', t_2 \rangle. \qquad (1.4.10)$$

We verify this by a more convenient reparameterization of the t integral, essentially identical with the parameterization in Section 4.10 of *Basic Matters*, for which purpose we introduce an integration parameter τ that ranges from $\tau = 0$ to $\tau = 1$, and regard $t, X(t), P(t)$ as functions of τ,

$$\tau = 0 : \quad t(\tau) = t_2,$$
$$\tau = 1 : \quad t(\tau) = t_1, \qquad (1.4.11)$$

where

$$dt = \frac{dt}{d\tau} d\tau \equiv \dot{t}\, d\tau \qquad (1.4.12)$$

and

$$\frac{dX}{dt} = \frac{dX}{d\tau} \frac{d\tau}{dt} \equiv \dot{X}/\dot{t} \qquad (1.4.13)$$

with dots denoting τ derivatives. Then

$$W_{12} = \int_0^1 d\tau \left(P\dot{X} - \dot{t}H \right), \qquad (1.4.14)$$

and variations of the "paths" $X(t), P(t)$ give

$$\delta W_{12} = \int_0^1 d\tau \left(\delta P \, \dot{X} + P \, \delta \dot{X} - \delta \dot{t} \, H - \dot{t} \, \delta H \right) \qquad (1.4.15)$$

where

$$\delta H = \delta H(X, P, t) = \frac{\partial H}{\partial X} \delta X + \delta P \frac{\partial H}{\partial P} + \delta t \frac{\partial H}{\partial t} \qquad (1.4.16)$$

or, after recalling the equations of motion (1.3.9),

$$\delta H = -\frac{dP}{dt} \delta X + \delta P \frac{dX}{dt} + \delta t \frac{\partial H}{\partial t} . \qquad (1.4.17)$$

Thus,

$$\delta W_{12} = \int_0^1 d\tau \left(\delta P \left(\dot{X} - \dot{t} \frac{dX}{dt} \right) + \left(P \, \delta \dot{X} + \dot{t} \frac{dP}{dt} \delta X \right) \right.$$
$$\left. - \left(\delta \dot{t} \, H + \dot{t} \, \delta t \frac{\partial H}{\partial t} \right) \right) . \qquad (1.4.18)$$

Here, the first term vanishes because $\dot{t} \dfrac{dX}{dt} = \dot{X}$, and we have

$$P \, \delta \dot{X} + \dot{t} \frac{dP}{dt} \delta X = P \, \delta \dot{X} + \dot{P} \, \delta X$$
$$= \frac{d}{d\tau} (P \, \delta X) \qquad (1.4.19)$$

for the second term, and

$$\delta \dot{t} \, H + \dot{t} \, \delta t \frac{\partial H}{\partial t} = \delta \dot{t} \, H + \delta t \, \dot{t} \frac{dH}{dt}$$
$$= \delta \dot{t} \, H + \delta t \frac{dH}{d\tau}$$
$$= \frac{d}{d\tau} (\delta t \, H) \qquad (1.4.20)$$

for the third term. Taken together they give

$$\delta W_{12} = (P \, \delta X - H \, \delta t) \Big|_{\tau = 0}^{\tau = 1}$$
$$= (P \, \delta X - H \, \delta t) \Big|_{t = t_2}^{t = t_1} = G_1 - G_2 \qquad (1.4.21)$$

with

$$G = P\,\delta X - H\,\delta t\,,\qquad(1.4.22)$$

the generator of (1.4.1)–(1.4.5), indeed.

In arriving at this result, we paid little attention to the order of P and dX or P and δX. This is justified because eventually $\delta X \to \delta x$ or $\delta x'$, so that we only need to consider variations of $X(t)$ that are multiples of the identity, and then the order of multiplication is irrelevant. One can extend the treatment to slightly more general variations, but this is not so important for the sequel, as we shall mainly use the explicit differential statement (1.4.6).

In Sections 3.3 and 3.4 of *Simple Systems*, we have applications of the endpoint variations supplied by $G_1 - G_2$ in (1.4.6). We now supplement them by a simple example for the $\delta H(t)$ contribution.

1.4.1 An example: Constant force

As an illustrative example, we consider the motion under a constant force of strength F, for which

$$H = \frac{1}{2M}P^2 - FX\qquad(1.4.23)$$

is the Hamilton operator. We already know the time transformation function $\langle x, t_1 | x', t_2 \rangle$ for $F = 0$, see (1.3.18), so we can get its $F \neq 0$ form by considering small changes of F. Now

$$\delta_F H = -\delta F\, X(t)\qquad(1.4.24)$$

so that

$$\delta_F \langle x, t_1 | x', t_2 \rangle = \frac{i}{\hbar}\delta F \langle x, t_1 | \int_{t_2}^{t_1} dt\, X(t) | x', t_2 \rangle\,,\qquad(1.4.25)$$

where we need $X(t)$ in terms of $X(t_1)$ and $X(t_2)$. The Heisenberg equations of motion

$$\frac{d}{dt}X(t) = \frac{1}{M}P(t)\,,\quad \frac{d}{dt}P(t) = F\qquad(1.4.26)$$

imply

$$X(t) = X(t_1)\frac{t - t_2}{T} + X(t_2)\frac{t_1 - t}{T} - \frac{F}{2M}(t_1 - t)(t - t_2)\qquad(1.4.27)$$

as one verifies by inspection. Accordingly,

$$\delta_F \langle x, t_1 | x', t_2 \rangle = \frac{i}{\hbar} \delta F \langle x, t_1 | x', t_2 \rangle \qquad (1.4.28)$$

$$\times \int_{t_2}^{t_1} dt \left[x \frac{t - t_2}{T} + x' \frac{t_1 - t}{T} - \frac{F}{2M}(t_1 - t)(t - t_2) \right]$$

or

$$\delta_F \log \langle x, t_1 | x', t_2 \rangle = \frac{\delta_F \langle x, t_1 | x', t_2 \rangle}{\langle x, t_1 | x', t_2 \rangle}$$

$$= \frac{i}{\hbar} \delta F \left(\frac{x + x'}{2} T - \frac{FT^3}{12M} \right) \qquad (1.4.29)$$

after making use of

$$\int_{t_2}^{t_1} dt \frac{t - t_2}{T} = \frac{1}{2} T = \int_{t_2}^{t_1} dt \frac{t_1 - t}{T} \qquad (1.4.30)$$

and

$$\int_{t_2}^{t_1} dt \, (t_1 - t)(t - t_2) = \frac{1}{6} T^3 . \qquad (1.4.31)$$

We recognize immediately that the right-hand side of (1.4.29) is a total variation in F,

$$\delta_F \log \langle x, t_1 | x', t_2 \rangle = \delta_F \left(\frac{i}{\hbar} \frac{x + x'}{2} FT - \frac{i}{\hbar} \frac{F^2 T^3}{24M} \right) , \qquad (1.4.32)$$

which implies

$$\langle x, t_1 | x', t_2 \rangle = \langle x, t_1 | x', t_2 \rangle \Big|_{F=0} e^{\frac{i}{\hbar} \frac{x+x'}{2} FT - \frac{i}{\hbar} \frac{F^2 T^3}{24M}} \qquad (1.4.33)$$

and we arrive at the time transformation function

$$\langle x, t_1 | x', t_2 \rangle = \sqrt{\frac{M}{i 2\pi \hbar T}} \, e^{\frac{i}{\hbar} \frac{M}{2T}(x - x')^2 + \frac{i}{\hbar} \frac{x+x'}{2} FT - \frac{i}{\hbar} \frac{F^2 T^3}{24M}} , \qquad (1.4.34)$$

in agreement with the result of Exercise 3-8 on page 67 in *Simple Systems*.

1-20 Repeat this for $\langle x, t_1 | p, t_2 \rangle$.

1-21 Consider the Hamilton operator

$$H = \frac{1}{2M}\left(P - \frac{\partial\lambda(X,t)}{\partial X}\right)^2 - \frac{\partial\lambda(X,t)}{\partial t}\,,$$

where $\lambda(X,t)$ is an arbitrary "gauge function" that depends on position operator X and parametrically on time t. Does the force $M\dfrac{\mathrm{d}^2}{\mathrm{d}t^2}X$ depend on λ? Find the λ dependence of $\langle x,t_1|x',t_2\rangle$.

1-22 Consider the Hamilton operator

$$H = \frac{1}{2M}P^2 + \frac{1}{2}\gamma(XP + PX)$$

with rate constant γ. Show that $\dfrac{\mathrm{d}}{\mathrm{d}t}(XP + PX) = \dfrac{2}{M}P^2$ and use this to find $X(t)P(t) + P(t)X(t)$ in terms of $X(t_1)$ and $P(t_2)$. Then employ the quantum action principle to determine first $\delta_\gamma\langle x,t_1|p,t_2\rangle$ and then $\langle x,t_1|p,t_2\rangle$.

1.4.2 Insertion: Varying an exponential function

As a preparation for the sequel, we derive an important mathematical formula for the response of e^A to infinitesimal variations of operator A. Begin with

$$\begin{aligned}
\delta\,\mathrm{e}^A &= \delta\sum_{k=0}^{\infty}\frac{1}{k!}A^k = \delta\sum_{k=1}^{\infty}\frac{1}{k!}A^k \\
&= \delta\sum_{k=0}^{\infty}\frac{1}{(k+1)!}A^{k+1} \\
&= \sum_{k=0}^{\infty}\frac{1}{(k+1)!}\delta A^{k+1}
\end{aligned} \tag{1.4.35}$$

where

$$\begin{aligned}
\delta A^{k+1} &= \delta A\,A^k + A\,\delta A\,A^{k-1} + A^2\,\delta A\,A^{k-2} + \cdots + A^k\,\delta A \\
&= \sum_{j=0}^{k}A^j\,\delta A\,A^{k-j}\,.
\end{aligned} \tag{1.4.36}$$

Therefore

$$\delta\,e^A = \sum_{k=0}^{\infty} \frac{1}{(k+1)!} \sum_{j=0}^{k} A^j\,\delta A\,A^{k-j}\,, \qquad (1.4.37)$$

or after rearranging the double sum

$$\delta\,e^A = \sum_{j,k=0}^{\infty} \frac{1}{(j+k+1)!} A^j\,\delta A\,A^k\,. \qquad (1.4.38)$$

With Leonhard Euler's so-called *beta function integral* ,

$$\frac{j!\,k!}{(j+k+1)!} = \int_0^1 \mathrm{d}x\,x^j(1-x)^k$$

$$= \int_0^1 \mathrm{d}x\,(1-x)^j x^k\,, \qquad (1.4.39)$$

this becomes

$$\delta\,e^A = \sum_{j,k=0}^{\infty} \int_0^1 \mathrm{d}x\,x^j(1-x)^k \frac{A^j}{j!}\,\delta A\,\frac{A^k}{k!}$$

$$= \int_0^1 \mathrm{d}x \sum_{j=0}^{\infty} \frac{(xA)^j}{j!}\,\delta A \sum_{k=0}^{\infty} \frac{\big((1-x)A\big)^k}{k!} \qquad (1.4.40)$$

or

$$\delta\,e^A = \int_0^1 \mathrm{d}x\,e^{xA}\,\delta A\,e^{(1-x)A}$$

$$= \int_0^1 \mathrm{d}x\,e^{(1-x)A}\,\delta A\,e^{xA}\,. \qquad (1.4.41)$$

This formula for the variation of an exponential operator function is worth memorizing. It contains all of perturbation theory *in nuce*.

1-23 Show that (1.4.41) implies

$$\delta\,e^{\alpha A} = \int_0^{\infty} \mathrm{d}\alpha_1 \int_0^{\infty} \mathrm{d}\alpha_2\,\delta(\alpha_1+\alpha_2-\alpha)\,e^{\alpha_1 A}\,\delta A\,e^{\alpha_2 A}\,, \qquad (1.4.42)$$

where α is a real parameter that is not varied along with A.

1-24 Now use this and the identity

$$(\beta - A)^{-1} = \frac{1}{\beta - A} = \int_0^\infty d\alpha \, e^{-\alpha\beta} e^{\alpha A}$$

to establish

$$\delta \frac{1}{\beta - A} = \frac{1}{\beta - A} \delta A \frac{1}{\beta - A}.$$

Which restrictions apply to β to ensure the convergence of the integral?

1-25 Justify the statement

$$\delta X^{-1} = -X^{-1} \delta X \, X^{-1},$$

and then use it to derive the result of the preceding exercise directly, that is without invoking the integral expression for $(\beta - A)^{-1}$.

1-26 First show that

$$e^{-\epsilon B} e^A e^{\epsilon B} = e^{e^{-\epsilon B} A e^{\epsilon B}},$$

where A and B are operators and ϵ is a complex number, and then use this to demonstrate that

$$\left[e^A, B \right] = \int_0^1 dx \, e^{(1-x)A} [A, B] \, e^{xA}.$$

1-27 All eigenvalues of the hermitian operator A are positive. Verify that

$$\log A = \int_0^\infty d\alpha \left(\frac{1}{\alpha + 1} - \frac{1}{\alpha + A} \right) = \int_0^\infty d\beta \, \frac{e^{-\beta} - e^{-\beta A}}{\beta}$$

are two valid integral representations of $\log A$. Then consider an infinitesimal variation δA and establish

$$\delta \log A = \int_0^\infty d\alpha \, \frac{1}{\alpha + A} \delta A \, \frac{1}{\alpha + A}.$$

1.4.3 *Time-independent Hamilton operator*

As an application of (1.4.41), which shows the connection with the quantum action principle, we consider the situation of a time-independent Hamilton operator, that is

$$\frac{d}{dt} H(A(t), t) = 0 \quad \text{implying} \quad H = H(A(t)) \equiv H(t) \tag{1.4.43}$$

so that $\dfrac{\partial H}{\partial t} = 0$ and $H(t) = H(t_1) = H(t_2)$. Then

$$\langle a, t_1 | = \langle a, t_2 | \, e^{-iH(t_2)(t_1 - t_2)/\hbar} \tag{1.4.44}$$

and, for variations of the Hamilton operator only,

$$\delta \langle a, t_1 | b, t_2 \rangle = \langle a, t_2 | \left(\delta \, e^{-iH(t_2)T/\hbar} \right) | b, t_2 \rangle \tag{1.4.45}$$

with $T = t_1 - t_2$ as always. Here we meet

$$\delta \, e^{-iH(t_2)T/\hbar} = \int_0^1 dx \; e^{(1 - x)(-iH(t_2)T/\hbar)} \left(-\frac{i}{\hbar} \delta H(t_2) T \right)$$
$$\times \; e^{x(-iH(t_2)T/\hbar)} \tag{1.4.46}$$

or with $xT = t - t_2$, $(1 - x)T = t_1 - t$, $dx\,T = dt$,

$$\delta \, e^{-iH(t_2)T/\hbar} = \int_{t_2}^{t_1} dt \; e^{-\frac{i}{\hbar} H(t_2)(t_1 - t)} \left(-\frac{i}{\hbar} \delta H(t_2) \right) e^{-\frac{i}{\hbar} H(t_2)(t - t_2)} \, . \tag{1.4.47}$$

The unitary operator on the far right, $e^{-\frac{i}{\hbar} H(t_2)(t - t_2)}$, advances states and operators from time t_2 to time t, so that

$$e^{\frac{i}{\hbar} H(t_2)(t - t_2)} \delta H(t_2) \, e^{-\frac{i}{\hbar} H(t_2)(t - t_2)} = \delta H(t) \tag{1.4.48}$$

and, therefore,

$$\delta \, e^{-iH(t_2)T/\hbar} = \int_{t_2}^{t_1} dt \; e^{-iH(t_2)(t_1 - t_2)/\hbar} \left(-\frac{i}{\hbar} \delta H(t) \right)$$
$$= e^{-iH(t_2)T/\hbar} \int_{t_2}^{t_1} dt \left(-\frac{i}{\hbar} \delta H(t) \right) . \tag{1.4.49}$$

It follows that

$$\delta \langle a, t_1 | b, t_2 \rangle = \langle a, t_2 | \, e^{-iH(t_2)T} \int_{t_2}^{t_1} dt \left(-\frac{i}{\hbar} \delta H(t) \right) | b, t_2 \rangle$$
$$= \langle a, t_1 | \int_{t_2}^{t_1} dt \left(-\frac{i}{\hbar} \delta H(t) \right) | b, t_2 \rangle , \tag{1.4.50}$$

which is exactly what the quantum action principle tells us about variations of the dynamics at intermediate times.

Chapter 2

Time-Dependent Perturbations

2.1 Born series

As another application of (1.4.41), we now consider the typical situation of a small perturbation, that is

$$H = H_0 + H_1 \,, \quad \frac{\partial H}{\partial t} = 0 \,, \tag{2.1.1}$$

where H_0 is the dominating part that governs the evolution mostly and H_1 is a perturbation that is small in some sense. It should then be a good approximation to take H_1 only into account to first, or perhaps second, order.

We regard the *unitary evolution operator*

$$e^{-iHT/\hbar} = e^{-i(H_0 + \lambda H_1)T/\hbar} \,, \tag{2.1.2}$$

where $\lambda = 1$, as a formal function of λ, which we expand in powers of λ,

$$
\begin{aligned}
e^{-iHT/\hbar} = \; & e^{-iH_0T/\hbar} \\
& + \lambda \left(\frac{\partial}{\partial \lambda} e^{-i(H_0 + \lambda H_1)T/\hbar} \bigg|_{\lambda=0} \right) \\
& + \frac{1}{2} \lambda^2 \left(\frac{\partial^2}{\partial \lambda^2} e^{-i(H_0 + \lambda H_1)T/\hbar} \bigg|_{\lambda=0} \right) \\
& + \cdots
\end{aligned}
\tag{2.1.3}
$$

with $\lambda = 1$ eventually. Thus

$$e^{-iHT/\hbar} = \underbrace{e^{-iH_0 T/\hbar}}_{\text{0th-order}}$$

$$+ \underbrace{\frac{\partial}{\partial \lambda} e^{-i(H_0 + \lambda H_1)T/\hbar}\bigg|_{\lambda=0}}_{\text{1st-order}}$$

$$+ \underbrace{\frac{1}{2}\left(\frac{\partial}{\partial \lambda}\right)^2 e^{-i(H_0 + \lambda H_1)T/\hbar}\bigg|_{\lambda=0}}_{\text{2nd-order}}$$

$$+ \cdots \tag{2.1.4}$$

which amounts to an expansion in powers of H_1, because λ always appears together with H_1.

The 1st-order term is readily available as an application of (1.4.41) or (1.4.42),

$$\frac{\partial}{\partial \lambda} e^{-i(H_0 + \lambda H_1)T/\hbar}$$

$$= \int_0^\infty dt_1 \int_0^\infty dt_2\, \delta(t_1 + t_2 - T)\, e^{-i(H_0 + \lambda H_1)t_1/\hbar}$$

$$\times \left(-\frac{i}{\hbar} H_1\right) e^{-i(H_0 + \lambda H_1)t_2/\hbar} \tag{2.1.5}$$

so that

$$\frac{\partial}{\partial \lambda} e^{-i(H_0 + \lambda H_1)T/\hbar}\bigg|_{\lambda=0}$$

$$= \int_0^\infty dt_1 \int_0^\infty dt_2\, \delta(t_1 + t_2 - T)\, e^{-iH_0 t_1/\hbar}\left(-\frac{i}{\hbar} H_1\right) e^{-iH_0 t_2/\hbar}. \tag{2.1.6}$$

Graphically, we can represent the 0th-order term by a straight line from $t = 0$ to $t = T$,

evolution under H_0 for the whole time span

and the 1st-order term is

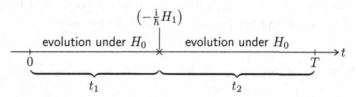

whereby the $\delta(t_1 + t_2 - T)$ factor in (2.1.6) enforces $t_1 + t_2 = T$.
The 2nd-order term is expected to give

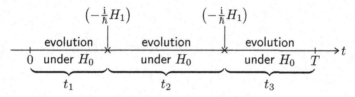

corresponding to

$$\frac{1}{2}\left(\frac{\partial}{\partial\lambda}\right)^2 e^{-i(H_0 + \lambda H_1)T/\hbar}\bigg|_{\lambda=0}$$

$$= \int_0^\infty dt_1 dt_2 dt_3 \, \delta(t_1 + t_2 + t_3 - T)\, e^{-iH_0 t_1/\hbar}\left(-\frac{i}{\hbar}H_1\right)$$

$$\times\, e^{-iH_0 t_2/\hbar}\left(-\frac{i}{\hbar}H_1\right) e^{-iH_0 t_3/\hbar} \qquad (2.1.7)$$

so that $T = t_1 + t_2 + t_3$ is broken up into three intervals of evolution under
H_0 with two applications of H_1 between these periods.

2-1 Derive this expression for the 2nd-order term.

We can continue like this and get the 3rd-, 4th-, 5th-, ... order terms
but, even without going through the technical steps, it is clear what we will
get eventually. Namely, the terms have this structure:

with, for the nth-order term, n applications of $\left(-\dfrac{i}{\hbar}H_1\right)$ and $n+1$ intervals
of evolution under H_0.

We have here, in (2.1.4) with (2.1.6) and (2.1.7), a first example of a
Born series, named after Max Born, for which we shall meet other examples

in Sections 2.2, 2.8, and 3.4.2. We regard as the defining property of a Born series that the successive terms have the structure exemplified by (2.1.6) and (2.1.7), which can be represented graphically as above.

2.2 Scattering operator

We exhibit the difference between the unitary evolution operator for H_0,

$$e^{-iH_0T/\hbar} \equiv U_0(T), \qquad (2.2.1)$$

and the one for $H = H_0 + H_1$,

$$e^{-iHT/\hbar} = e^{-i(H_0 + H_1)T/\hbar} \equiv U(T), \qquad (2.2.2)$$

by introducing the *scattering operator* $S(T)$,

$$U(T) = U_0(T)S(T). \qquad (2.2.3)$$

For $H_1 \equiv 0$, we have $S(T) = 1$, of course, so that $S(T)$ summarizes the net effect of H_1 in the sense that $S(T)$ accounts for the change in the evolution that originates in H_1. The unperturbed evolution governed by H_0 is put aside by extracting the factor $U_0(T)$ out of $U(T)$.

The name "scattering operator" alludes to a frequent application, namely in scattering theory. Generally speaking, the system evolving under $H = H_0 + H_1$ is perturbed by H_1 and, therefore, does not follow the trajectory laid out by H_0. Rather it is "scattered" off that path by the action of H_1. It should be clear that systematic expansions in powers of H_1, such as the Born series, have as a prerequisite the requirement that H_1 is small.

The Born series for $U(T)$ can be converted into a corresponding series for $S(T)$,

$$
\begin{aligned}
S(T) &= U_0(T)^{-1}U(T) = U_0(-T)U(T) \\
&= U_0(T)^{-1}\left[U_0(T) - \frac{i}{\hbar}\int_0^T dt\, U_0(T-t)H_1U_0(t) \right.\\
&\quad + \left(-\frac{i}{\hbar}\right)^2 \int_0^T dt \int_0^t dt'\, U_0(T-t)H_1U_0(t-t')H_1U_0(t') \\
&\quad \left. + \cdots \right]
\end{aligned}
\qquad (2.2.4)
$$

or

$$S(T) = 1 - \frac{i}{\hbar} \int_0^T dt\, U_0(-t) H_1 U_0(t)$$

$$+ \left(-\frac{i}{\hbar}\right)^2 \int_0^T dt \int_0^t dt'\, U_0(-t) H_1 U_0(t) U_0(-t') H_1 U_0(t')$$

$$+ \cdots, \tag{2.2.5}$$

after we made use of the unitary property $U_0(T)^{-1} = U_0(T)^\dagger = U_0(-T)$ and the group property $U_0(t_1 + t_2) = U_0(t_1)U_0(t_2)$. All H_1 appearing here are sandwiched by a $U_0(-t), U_0(t)$ pair of evolution operators, which invites us to introduce

$$\overline{H_1}(t) \equiv U_0(-t) H_1 U_0(t) \tag{2.2.6}$$

as a convenient abbreviation. This is often referred to as "H_1 in the interaction picture" and so be it, but since the standard notions of "Schrödinger picture", "Heisenberg picture", and "interaction picture", sometimes also called "Dirac picture", are more confusing than enlightening, we shall not employ such terminology.

Then

$$S(T) = 1 - \frac{i}{\hbar} \int_0^T dt\, \overline{H_1}(t)$$

$$+ \left(-\frac{i}{\hbar}\right)^2 \int_0^T dt \int_0^t dt'\, \overline{H_1}(t)\overline{H_1}(t')$$

$$+ \cdots, \tag{2.2.7}$$

is the Born series for the scattering operator, whereby the kth term would have k factors of $\overline{H_1}(t)$ with different time arguments such that the $\overline{H_1}$s for later times stand to the left of the ones for earlier times; more about this in Section 2.3.

We note that the Born series has a self-repeating pattern,

$$S(T) = 1 - \frac{i}{\hbar} \int_0^T dt\, \overline{H_1}(t)$$

$$\times \left[1 - \frac{i}{\hbar} \int_0^t dt'\, \overline{H_1}(t') \right.$$

$$\left. + \left(-\frac{i}{\hbar}\right)^2 \int_0^t dt' \int_0^{t'} dt''\, \overline{H_1}(t')\overline{H_1}(t'') + \cdots \right], \tag{2.2.8}$$

where the series in the square brackets is the Born series again, for $S(t)$,

$$\left[1 - \frac{i}{\hbar} \int_0^t dt' \, \overline{H_1}(t') + \cdots \right] = S(t) \,. \tag{2.2.9}$$

As a consequence, we have an integral equation for the scattering operator $S(T)$,

$$S(T) = 1 - \frac{i}{\hbar} \int_0^T dt \, \overline{H_1}(t) S(t) \,, \tag{2.2.10}$$

which has a structure that is met rather often in perturbation theory, the structure of a *Lippmann–Schwinger equation*, named after Bernard A. Lippmann and Julian Schwinger. It can be used for systematic iterations in the form

$$S_{n+1}(T) = 1 - \frac{i}{\hbar} \int_0^T dt \, \overline{H_1}(t) S_n(t) \tag{2.2.11}$$

where, it is hoped, S_{n+1} is an improvement over S_n and $S_n \to S$ as $n \to \infty$.

2-2 Begin with $S_0(t) \equiv 1$ and find $S_1(T)$ as well as $S_2(T)$. Compare with the Born series in (2.2.7).

2-3 The so-called *reaction operator* $K(T)$ is defined by

$$S = \frac{1 - \frac{i}{2}K}{1 + \frac{i}{2}K} \,.$$

Show that K is hermitian. What is the 1st-order Born approximation for K? What about the 2nd-order approximation?

2.3 Dyson series

The 2nd-order term in the Born series (2.2.7) for the scattering operator,

$$\left(-\frac{i}{\hbar}\right)^2 \int_0^T dt \int_0^t dt' \, \overline{H_1}(t) \overline{H_1}(t') \,, \tag{2.3.1}$$

can also be written as

$$\left(-\frac{i}{\hbar}\right)^2 \int_0^T dt' \int_0^{t'} dt \, \overline{H_1}(t') \overline{H_1}(t) = \left(-\frac{i}{\hbar}\right)^2 \int_0^T dt \int_t^T dt' \, \overline{H_1}(t') \overline{H_1}(t) \tag{2.3.2}$$

upon interchanging the integration variables $t \leftrightarrow t'$. We combine them into half the sum of both,

$$\frac{1}{2}\left(-\frac{i}{\hbar}\right)^2 \int_0^T dt \int_0^T dt' \left[\overline{H_1}(t)\overline{H_1}(t')\right]\Bigg|_{\substack{\text{later time to the left} \\ \text{of the earlier time}}} \tag{2.3.3}$$

where we have this injunction about *time ordering*. We write $[\cdots]_+$ for such a time-ordered product of $\overline{H_1}$ factors, as illustrated by

$$\left[\overline{H_1}(t)\overline{H_1}(t')\right]_+ = \begin{cases} \overline{H_1}(t)\overline{H_1}(t') & \text{if } t > t', \\[2mm] \overline{H_1}(t')\overline{H_1}(t) & \text{if } t' > t, \end{cases} \tag{2.3.4}$$

for two factors as above, and by

$$\left[\overline{H_1}(t)\overline{H_1}(t')\overline{H_1}(t'')\right]_+ = \begin{cases} \overline{H_1}(t)\overline{H_1}(t')\overline{H_1}(t'') & \text{if } t > t' > t'', \\[2mm] \overline{H_1}(t)\overline{H_1}(t'')\overline{H_1}(t') & \text{if } t > t'' > t', \\[2mm] \quad\vdots \\[2mm] \overline{H_1}(t'')\overline{H_1}(t')\overline{H_1}(t) & \text{if } t'' > t' > t, \end{cases} \tag{2.3.5}$$

for three factors, with altogether $6 = 3!$ arrangements. So, the 3rd-order term in the Born series is

$$\left(-\frac{i}{\hbar}\right)^3 \int_0^T dt \int_0^t dt' \int_0^{t'} dt'' \, \overline{H_1}(t)\overline{H_1}(t')\overline{H_1}(t'')$$

$$= \frac{1}{3!}\left(-\frac{i}{\hbar}\right)^3 \int_0^T dt \int_0^T dt' \int_0^T dt'' \left[\overline{H_1}(t)\overline{H_1}(t')\overline{H_1}(t'')\right]_+ \tag{2.3.6}$$

for which

$$\frac{1}{3!}\left[\left(-\frac{i}{\hbar}\int_0^T dt\, \overline{H_1}(t)\right)^3\right]_+ \tag{2.3.7}$$

is a suggestive, compact notation. Likewise, we get

$$\frac{1}{k!}\left[\left(-\frac{i}{\hbar}\int_0^T dt\, \overline{H_1}(t)\right)^k\right]_+ \tag{2.3.8}$$

for the kth-order term, and then

$$S(T) = \left[\sum_{k=0}^{\infty} \frac{1}{k!} \left(-\frac{i}{\hbar} \int_0^T dt \, \overline{H_1}(t) \right)^k \right]_+$$
$$= \left[\exp\left(-\frac{i}{\hbar} \int_0^T dt \, \overline{H_1}(t) \right) \right]_+ \qquad (2.3.9)$$

is a *formal* summation of the Born series in terms of a time-ordered exponential function. Of course, this is just a compact way of writing the original Born series (2.2.7), but nevertheless it represents a very important advance of the formalism because many formal manipulations are facilitated much by this notation. One refers to this version of the Born series as the *Dyson series*, named after Freeman J. Dyson.

2-4 What becomes of the Dyson series when H_1 commutes with H_0? Why is this case of little interest?

2-5 For $H_0 = \frac{1}{2M} P^2$ and $H_1 = -FX$, what is $\overline{H_1}(t)$?

2-6 Calculate the 1st- and 2nd-order terms of the Born series for this example.

2.4 Fermi's golden rule

A typical question is this: Given that the system is initially in an eigenstate of H_0, what is the chance that H_1 will effect a transition to another eigenstate of H_0? The answer is given, under general and typical circumstances, by what is arguably the most famous and most frequently applied statement of perturbation theory, Enrico Fermi's celebrated *golden rule*. It is an application of the Born series, or the Dyson series, to the lowest order, that is first order, in the perturbation H_1.

Thus, we have eigenket $|n\rangle$ of H_0 initially,

$$H_0 |n\rangle = |n\rangle E_n , \qquad (2.4.1)$$

and eigenbra $\langle m|$ of H_0 finally,

$$\langle m| H_0 = E_m \langle m| . \qquad (2.4.2)$$

The transition probability is

$$\left|\langle m|U(T)|n\rangle\right|^2 = \left|\langle m|U_0(T)S(T)|n\rangle\right|^2$$
$$= \left|e^{-\frac{i}{\hbar}E_m T}\langle m|S(T)|n\rangle\right|^2$$
$$= \left|\langle m|S(T)|n\rangle\right|^2 \tag{2.4.3}$$

where, to first order in H_1,

$$\langle m|S(T)|n\rangle \cong \langle m|\left[1 - \frac{i}{\hbar}\int_0^T dt\,\overline{H_1}(t)\right]|n\rangle$$
$$= \delta_{mn} - \frac{i}{\hbar}\int_0^T dt\,\langle m|\overline{H_1}(t)|n\rangle . \tag{2.4.4}$$

Now, we have a *transition* in mind, so that the final state is different from the initial state, $m \neq n$, implying $\delta_{mn} \to 0$. Further, recall how $\overline{H_1}(t)$ is related to H_1 through the sandwiching with $U_0(-t)$ and $U_0(t)$,

$$\langle m|\overline{H_1}(t)|n\rangle = \langle m|U_0(-t)H_1 U_0(t)|n\rangle$$
$$= e^{\frac{i}{\hbar}E_m t}\langle m|H_1|n\rangle\,e^{-\frac{i}{\hbar}E_n t}$$
$$= e^{\frac{i}{\hbar}(E_m - E_n)t}\langle m|H_1|n\rangle . \tag{2.4.5}$$

Accordingly,

$$\left|\langle m|U(T)|n\rangle\right|^2 \cong \left|-\frac{i}{\hbar}\langle m|H_1|n\rangle\right|^2\left|\int_0^T dt\,e^{\frac{i}{\hbar}(E_m - E_n)t}\right|^2$$
$$= \frac{1}{\hbar^2}\left|\langle m|H_1|n\rangle\right|^2\left|\int_0^T dt\,e^{-i\omega t}\right|^2 \tag{2.4.6}$$

where

$$\omega = \frac{1}{\hbar}(E_n - E_m) \tag{2.4.7}$$

is the *transition frequency*, the energy difference (of the eigenstates of H_0, not of H!) expressed as a circular frequency. The integral is

$$\int_0^T dt \, e^{i\omega t} = \frac{e^{-i\omega T} - 1}{-i\omega}$$

$$= \frac{e^{-i\omega T/2}}{-i\omega} \underbrace{\left(e^{-i\omega T/2} - e^{i\omega T/2} \right)}_{= \, -2i\sin\frac{\omega T}{2}}$$

$$= e^{-i\omega T/2} \frac{2}{\omega} \sin\frac{\omega T}{2} \,, \tag{2.4.8}$$

where the first factor has unit modulus, so that

$$\left| \langle m|U(T)|n\rangle \right|^2 \cong \frac{1}{\hbar^2} \left| \langle m|H_1|n\rangle \right|^2 \left(\frac{2}{\omega} \sin\frac{\omega T}{2} \right)^2 . \tag{2.4.9}$$

Before proceeding we must think a bit about the situation we are envisioning. The Hamilton operator $H = H_0 + H_1$ is dominated by H_0 and we take the small perturbation into account only to lowest order. Therefore, consistency requires that we stick to time scales that are long on the scale set by the H_0 transition frequency ω but not long in an absolute sense, $T \not\to \infty$. In short, while $\omega T \gg 1$, the duration T is a finite time so small that the net effect of H_1 is still small, which is to say that the probability of the transition $n \to m$ is a small number. It should, therefore, be possible to speak of a *transition rate* γ,

$$\left| \langle m|U(T)|n\rangle \right|^2 \cong \gamma T \ll 1 \,. \tag{2.4.10}$$

We are thus asked to identify the regime in which $\left| \langle m|U(T)|n\rangle \right|^2$ grows linearly with time T. This is *not* the regime of very short times because we know (see Section 5.1.3 in *Basic Matters*) that the persistence probability deviates by an amount $\propto T^2$ from unity when the elapsed time T is small,

$$\underbrace{\left| \langle n|U(T)|n\rangle \right|^2}_{\substack{\text{same state} \\ \text{initially and finally}}} \cong 1 - \left(\frac{T}{\hbar}\delta H \right)^2 \quad \text{for very small } T \,, \tag{2.4.11}$$

implying that $\left| \langle m|U(T)|n\rangle \right|^2 \propto T^2$ for $n \neq m$. The energy spread δH that appears here is the spread of $H = H_0 + H_1$,

$$\delta H = \sqrt{\langle n|H^2|n\rangle - \langle n|H|n\rangle^2} \,, \tag{2.4.12}$$

but since $|n\rangle$ is an eigenket of H_0, it is also the spread of H_1,

$$\delta H = \delta H_1 = \sqrt{\langle n|H_1^2|n\rangle - \langle n|H_1|n\rangle^2}\,.\tag{2.4.13}$$

2-7 Verify this, and relate the general short-time form (2.4.11) to the small-T version of $\left|\langle m|U(T)|n\rangle\right|^2$ that you get from (2.4.9).

In summary, it is $\omega T \gg 1$ that we must consider more closely. We note a mathematical fact, namely

$$\frac{\left(\sin(x/\epsilon)\right)^2}{x^2/\epsilon} \to \pi\delta(x) \quad \text{as } \epsilon \to 0\,,\tag{2.4.14}$$

for which we give the usual plausibility argument. A plot of $\dfrac{\left(\sin(x/\epsilon)\right)^2}{x^2/\epsilon}$ has this general appearance:

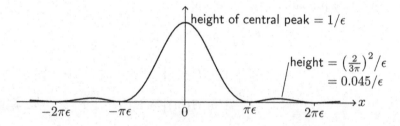

As $\epsilon \to 0$, the central peak only has area worth speaking of, and it is a very narrow, very tall peak. The area is

$$\int_{-\infty}^{\infty} \mathrm{d}x\, \frac{\left(\sin(x/\epsilon)\right)^2}{x^2/\epsilon} = \pi\tag{2.4.15}$$

as one shows easily by an integration by parts that turns this integral into the well-known integral

$$\int_{-\infty}^{\infty} \mathrm{d}x\, \frac{\sin x}{x} = \pi\,.\tag{2.4.16}$$

Actually, there is a little swindle in this argument because the area of the central peak is 90.3% of the total area, independent of ϵ, that of the first side peaks (on both sides) is 4.7%, and the second side peaks have 1.7% of the area. A better argument would thus exploit that, as $\epsilon \to 0$, the region near $x = 0$ contains more and more peaks and the total area grows to unity for any fixed interval around $x = 0$.

2-8 Use the general procedure of (4.1.18) in *Basic Matters* for

$$D(\phi) = \begin{cases} 1 - \frac{1}{2}|\phi| & \text{for} \quad |\phi| < 2, \\ 0 & \text{for} \quad |\phi| > 2, \end{cases}$$

to give an alternative, and perhaps more convincing, derivation of (2.4.14).

Having thus established (2.4.14), we now apply it to

$$\epsilon = \frac{2}{T} \quad \text{and} \quad x = \omega, \tag{2.4.17}$$

meaning

$$\left(\frac{2}{\omega}\sin\frac{\omega T}{2}\right)^2 = 2T\frac{\left(\sin\frac{\omega T}{2}\right)^2}{\frac{\omega^2 T}{2}}$$

$$\cong 2\pi T\delta(\omega) \quad \text{for large } T. \tag{2.4.18}$$

The transition probability is, therefore,

$$\begin{aligned}
\left|\langle m|U(T)|n\rangle\right|^2 &\cong \frac{1}{\hbar^2}\left|\langle m|H_1|n\rangle\right|^2 2\pi T\delta(\omega) \\
&= \frac{2\pi}{\hbar}\left|\langle m|H_1|n\rangle\right|^2\frac{1}{\hbar}\delta\left(\frac{E_n - E_m}{\hbar}\right)T \\
&= \frac{2\pi}{\hbar}\left|\langle m|H_1|n\rangle\right|^2\delta(E_n - E_m)T, \tag{2.4.19}
\end{aligned}$$

and the transition rate γ of (2.4.10) is then

$$\gamma_{n\to m} \equiv \frac{2\pi}{\hbar}\left|\langle m|H_1|n\rangle\right|^2\delta(E_n - E_m). \tag{2.4.20}$$

This is Fermi's *golden rule*.

It is important to realize that the rate depends on both the initial and the final state, which we emphasize by writing $\gamma_{n\to m}$. But, in the typical situation in which this applies, there are many final states that we cannot usually distinguish from each other. Say, for instance, we talk about an atom that emits a photon, thereby undergoing a transition from an excited to a ground state. The photon is emitted into a range of frequencies and into many directions. Then there are a lot of final states, all with the same final energy $E' = E_m$ but many different values for the quantum numbers symbolized by subscript m. We must average $\left|\langle m|H_1|n\rangle\right|^2$ over all those states,

$$\left|\langle m|H_1|n\rangle\right|^2 \longrightarrow \overline{\left|\langle m|H_1|n\rangle\right|^2}, \tag{2.4.21}$$

and multiply by the density $\rho(E')$ of the final states, followed by integrating over E'. Together, this gives the transition rate

$$\gamma = \int dE' \, \rho(E') \frac{2\pi}{\hbar} \overline{\left|\langle m|H_1|n\rangle\right|^2} \delta(E_n - E_m)$$
$$= \frac{2\pi}{\hbar} \overline{\left|\langle m|H_1|n\rangle\right|^2} \rho(E_n) \,. \tag{2.4.22}$$

The averaging $\overline{\langle \cdots \rangle^2}$ is, as said above, over all final states with $E' = E_m$ ($\equiv E_n$ by virtue of the δ function), and $\rho(E)$ is the density of these final states, normalized in accordance with

$$\int dE' \, \rho(E') = 1 \,. \tag{2.4.23}$$

2.5 Photon emission by a "two-level atom"

2.5.1 *Golden-rule treatment*

As an application of the golden rule let us consider a simplified model of an atom emitting a photon. We reduce the atom to two energetic levels, the excited state $|e\rangle$ and the ground state $|g\rangle$, which are just the two atomic states involved in the transition:

initial state:
atom excited,
no photons

final state:
atom deexcited,
one photon

For the description of the atom we use the transition operators

$$\sigma = |g\rangle\langle e| \,, \quad \sigma^\dagger = |e\rangle\langle g| \tag{2.5.1}$$

whose products are the projectors on the atomic states,

$$\sigma^\dagger \sigma = |e\rangle\langle e| \,, \quad \sigma\sigma^\dagger = |g\rangle\langle g| \,. \tag{2.5.2}$$

Upon assigning energy $\hbar\omega$ to the excited state and energy 0 to the ground state, we thus have

$$H_{\text{atom}} = \hbar\omega\sigma^\dagger\sigma \tag{2.5.3}$$

for the atom Hamilton operator.

The photons are a collection of harmonic oscillators with energy steps of $\hbar\omega_\nu$ for the νth oscillator, that is the νth kind of photon, and oscillator ladder operators $A_\nu, A_\nu^\dagger$, one for each photon kind. The label ν stands for all characterizing properties of the νth kind, such as propagation direction and polarization. Then

$$H_{\text{phot}} = \sum_\nu \hbar\omega_\nu A_\nu^\dagger A_\nu \qquad (2.5.4)$$

is the photon part of the Hamilton operator.

The interaction is modeled in the simplest form, namely by just assuming the obvious: when the atom undergoes $e \to g$ one photon is emitted, and the transition $g \to e$ is accompanied by the absorption of one photon. The basic operators are, therefore,

$$\sigma A_\nu^\dagger \quad \text{for} \qquad \text{transition } e \to g \text{ and photon emitted,} \qquad (2.5.5)$$

and

$$\sigma^\dagger A_\nu \quad \text{for} \qquad \text{transition } g \to e \text{ and photon absorbed,} \qquad (2.5.6)$$

when $\sigma A_\nu^\dagger$ and $\sigma^\dagger A_\nu$ act on kets. The strength of these processes is measured by the so-called Rabi frequencies Ω_ν, named after Isidor I. Rabi, so that

$$H_{\text{int}} = -\sum_\nu \hbar\left(\Omega_\nu \sigma A_\nu^\dagger + \Omega_\nu^* \sigma^\dagger A_\nu\right) \qquad (2.5.7)$$

is the interaction part of the Hamilton operator. The overall minus sign is conventional and of no physical significance.

Taken together we have

$$H = H_{\text{atom}} + H_{\text{photon}} + H_{\text{int}}$$
$$= \underbrace{\hbar\omega\sigma^\dagger\sigma + \sum_\nu \hbar\omega_\nu A_\nu^\dagger A_\nu}_{= H_0} - \underbrace{\sum_\nu \hbar\left(\Omega_\nu \sigma A_\nu^\dagger + \Omega_\nu^* \sigma^\dagger A_\nu\right)}_{= H_1} \qquad (2.5.8)$$

so that the eigenstates of H_0 are

$|n\rangle \cong |e, 0\rangle$ initially: atom excited, no photons;

$\langle m| \cong \langle g, 1_\nu|$ finally: atom deexcited, one photon of kind ν.   (2.5.9)

According to Fermi's golden rule, we thus have

$$\gamma_\nu = \frac{2\pi}{\hbar} \left| \langle g, 1_\nu | H_{\text{int}} | e, 0 \rangle \right|^2 \delta(\hbar\omega - \hbar\omega_\nu) \tag{2.5.10}$$

for the partial rate that goes with the emission of a photon of the νth kind. The argument of the δ function is the difference of the H_0 eigenvalues, namely $\hbar\omega$ for $|e, 0\rangle$ and $\hbar\omega_\nu$ for $\langle g, 1_\nu |$. The evaluation of

$$\langle g, 1_\nu | H_{\text{int}} | e, 0 \rangle = -\hbar \sum_{\nu'} \langle g, 1_\nu | \left(\Omega_{\nu'} \sigma A_{\nu'}^\dagger + \Omega_{\nu'}^* \sigma^\dagger A_{\nu'} \right) | e, 0 \rangle$$

$$= -\hbar\Omega_\nu \tag{2.5.11}$$

is straightforward because $\langle g | \sigma^\dagger = 0$ and

$$\langle g, 1_\nu | \sigma A_{\nu'}^\dagger = \delta_{\nu\nu'} \langle e, 0 |, \tag{2.5.12}$$

and there is only a single term contributing to the sum in (2.5.11). It follows that

$$\gamma_\nu = \frac{2\pi}{\hbar} |\hbar\Omega_\nu|^2 \frac{1}{\hbar} \delta(\omega - \omega_\nu) = 2\pi |\Omega_\nu|^2 \delta(\omega - \omega_\nu). \tag{2.5.13}$$

Next, we note that there are many photon kinds with the same frequency $\omega = \omega_\nu$, differing only in their polarization and propagation directions. For given frequency ω_ν, we average over those other attributes,

$$|\Omega_\nu|^2 \longrightarrow \overline{|\Omega_\nu|^2} \Big|_{\substack{\text{average over all} \\ \text{modes with } \omega_\nu = \omega'}}, \tag{2.5.14}$$

and multiply with the density $\rho(\omega')$ of photon modes ("mode" means "kind of photon" as labeled by ν). Then

$$\gamma = \int d\omega' \, 2\pi \overline{|\Omega_\nu|^2} \, \delta(\omega - \omega')\rho(\omega') = 2\pi f(\omega) \tag{2.5.15}$$

is the resulting expression for the transition rate, where

$$f(\omega') = \rho(\omega') \, \overline{|\Omega_\nu|^2} \Big|_{\text{av. for } \omega_\nu = \omega'} \tag{2.5.16}$$

summarizes the essential atomic properties specified by the Rabi frequencies Ω_ν and the relevant properties of the electromagnetic radiation field specified by the mode density $\rho(\omega')$. The former is to be derived in atomic theory, the latter in electromagnetic theory. For the purpose of the present discussion, we take for granted that $f(\omega')$ is a known function of ω'.

2.5.2   *A more detailed treatment*

The golden rule gives us the transition rate $\gamma = 2\pi f(\omega)$ but hardly any other detail. So let us try to do better and, following Victor F. Weisskopf and Eugene P. Wigner, attempt to solve the Schrödinger equation.

We begin with writing the state ket as a superposition of $|e, 0\rangle$ and $|g, 1_\nu\rangle$ at time t,

$$| \; \rangle = |e, 0, t\rangle \alpha(t) + \sum_\nu |g, 1_\nu, t\rangle \beta_\nu(t) , \tag{2.5.17}$$

with probability amplitudes

$$\begin{aligned} \alpha(t) &= \langle e, 0, t| \; \rangle , \\ \beta_\nu(t) &= \langle g, 1_\nu, t| \; \rangle . \end{aligned} \tag{2.5.18}$$

They obey their respective Schrödinger equations,

$$\begin{aligned} \mathrm{i}\hbar \frac{\partial}{\partial t}\alpha(t) &= \langle e, 0, t|H| \; \rangle \\ &= \hbar\omega \langle e, 0, t| \; \rangle - \hbar \sum_\nu \Omega_\nu^* \langle g, 1_\nu, t| \; \rangle \\ &= \hbar\omega\alpha(t) - \hbar \sum_\nu \Omega_\nu^* \beta_\nu(t) \end{aligned} \tag{2.5.19}$$

and

$$\begin{aligned} \mathrm{i}\hbar \frac{\partial}{\partial t}\beta_\nu(t) &= \langle g, 1_\nu, t|H| \; \rangle \\ &= \hbar\omega_\nu \langle g, 1_\nu, t| \; \rangle - \hbar\Omega_\nu \langle e, 0, t| \; \rangle \\ &= \hbar\omega_\nu\beta_\nu(t) - \hbar\Omega_\nu\alpha(t) , \end{aligned} \tag{2.5.20}$$

or, after removing the common factor of $\hbar$,

$$\begin{aligned} \mathrm{i}\frac{\partial}{\partial t}\alpha &= \omega\alpha - \sum_\nu \Omega_\nu^* \beta_\nu , \\ \mathrm{i}\frac{\partial}{\partial t}\beta_\nu &= \omega_\nu\beta_\nu - \Omega_\nu\alpha . \end{aligned} \tag{2.5.21}$$

This is a coupled system of very many equations as there are very many photon modes. We have to solve it subject to the initial conditions

$$\alpha(0) = 1 , \quad \beta_\nu(0) = 0 , \tag{2.5.22}$$

corresponding to the situation of "atom excited, no photon at $t = 0$". In the course of time, $\alpha(t)$ will decrease and $\beta_\nu(0)$ will increase but the total probability remains fixed at 100%, of course,

$$\underbrace{|\alpha(t)|^2}_{\substack{\text{probability} \\ \text{of } |e, 0\rangle \\ \text{at time } t}} + \sum_\nu \underbrace{|\beta_\nu(t)|^2}_{\substack{\text{probability} \\ \text{of } |g, 1_\nu\rangle \\ \text{at time } t}} = 1, \qquad (2.5.23)$$

as one verifies easily.

We can express $\beta_\nu(t)$ in terms of $\alpha(t)$ by solving the β_ν equation formally, beginning with

$$\left(i\frac{\partial}{\partial t} - \omega_\nu\right)\beta_\nu(t) = e^{-i\omega_\nu t}i\frac{\partial}{\partial t}\left(e^{i\omega_\nu t}\beta_\nu(t)\right)$$
$$= -\Omega_\nu \alpha(t) \qquad (2.5.24)$$

or

$$\frac{\partial}{\partial t}\left(e^{i\omega_\nu t}\beta_\nu(t)\right) = i\Omega_\nu\, e^{i\omega_\nu t}\alpha(t), \qquad (2.5.25)$$

leading to

$$\beta_\nu(t) = i\Omega_\nu \int_0^t dt'\, e^{-i\omega_\nu(t - t')}\alpha(t'), \qquad (2.5.26)$$

where $\beta_\nu(0) = 0$ is taken into account. We use this in the α equation of (2.5.21) for the elimination of the β_ν:

$$\left(i\frac{\partial}{\partial t} - \omega\right)\alpha(t) = e^{-i\omega t}i\frac{\partial}{\partial t}\left(e^{i\omega t}\alpha(t)\right)$$
$$= -\sum_\nu \Omega_\nu^* \beta_\nu(t)$$
$$= -i\sum_\nu |\Omega_\nu|^2 \int_0^t dt'\, e^{-i\omega_\nu(t - t')}\alpha(t'). \qquad (2.5.27)$$

It will be convenient to exhibit the product $e^{i\omega t}\alpha(t)$,

$$\frac{\partial}{\partial t}\left(e^{i\omega t}\alpha(t)\right) = -\sum_\nu |\Omega_\nu|^2 \int_0^t dt'\, e^{i(\omega - \omega_\nu)(t - t')}\left(e^{i\omega t'}\alpha(t')\right), \quad (2.5.28)$$

because we expect that $\alpha(t)$ is dominated by its natural oscillation with frequency ω, with a modification that results from the interaction part of the Hamilton operator. Indeed, if there were no interaction, that is: all $\Omega_\nu = 0$, the right-hand side would vanish.

On physical grounds we expect that $\alpha(t)$ obeys an exponential decay law, at least approximately and for the long, but not extremely long, times beyond the initial $\propto t^2$ decay period. Therefore, we try the *ansatz*

$$\alpha(t) \cong e^{-i\omega t} e^{-\frac{1}{2}\Gamma t} \tag{2.5.29}$$

where Γ is meant to be independent of time t. Clearly, this cannot be the exact solution of the differential-integral equation (2.5.28) for $\alpha(t)$, but let us try it nevertheless because it is a physically motivated approximation that is reasonably plausible.

On the left-hand side of (2.5.28) we get

$$\frac{\partial}{\partial t}\left(e^{i\omega t}\alpha(t)\right) \cong -\frac{1}{2}\Gamma\, e^{-\frac{1}{2}\Gamma t} \tag{2.5.30}$$

and on the right-hand side

$$
\begin{aligned}
&-\sum_\nu |\Omega_\nu|^2 \int_0^t dt'\, e^{i(\omega - \omega_\nu)(t - t')}\, e^{-\frac{1}{2}\Gamma t'} \\
&= -e^{-\frac{1}{2}\Gamma t} \sum_\nu |\Omega_\nu|^2 \int_0^t dt'\, e^{[i(\omega - \omega_\nu) + \frac{1}{2}\Gamma](t - t')} \\
&= -e^{-\frac{1}{2}\Gamma t} \sum_\nu |\Omega_\nu|^2 \left(-\frac{1 - e^{[i(\omega - \omega_\nu) + \frac{1}{2}\Gamma]t}}{i(\omega - \omega_\nu) + \frac{1}{2}\Gamma}\right)
\end{aligned}
\tag{2.5.31}
$$

so that

$$\frac{1}{2}\Gamma \cong i\sum_\nu |\Omega_\nu|^2 \frac{1 - e^{i(\omega - \omega_\nu)t}\, e^{\frac{1}{2}\Gamma t}}{\omega - \omega_\nu - \frac{i}{2}\Gamma}. \tag{2.5.32}$$

Now, just as in the step that took us from γ_ν to γ in Section 2.4 we apply the averaging over modes with common frequency $\omega' = \omega_\nu$ and introduce the mode density $\rho(\omega')$, as summarized in the rule

$$
\begin{aligned}
\sum_\nu |\Omega_\nu|^2 a(\omega_\nu) &\to \int d\omega'\, \rho(\omega')\overline{|\Omega_\nu|^2}\, a(\omega') \\
&= \int d\omega'\, f(\omega')a(\omega')
\end{aligned}
\tag{2.5.33}
$$

for the function $a(\omega')$ in question, here: the ratio in (2.5.32). Thus we arrive at

$$\frac{1}{2}\Gamma \cong i \int d\omega' \, f(\omega') \frac{1 - e^{-i(\omega' - \omega)t} \, e^{\frac{1}{2}\Gamma t}}{\omega - \omega' - \frac{i}{2}\Gamma} \, . \qquad (2.5.34)$$

The main contribution to this integral comes from the vicinity of $\omega' \cong \omega$ because Γ is small (otherwise the treatment would be inconsistent) and the integrand is, therefore, dominated by the $\omega' - \omega$ terms in its time dependence and size. So, we continue to be "brutal" in our approximate treatment and just neglect the Γ terms on the right,

$$\frac{1 - e^{-i(\omega' - \omega)t} \, e^{\frac{1}{2}\Gamma t}}{\omega - \omega' - \frac{i}{2}\Gamma} \rightarrow \frac{1 - e^{-i(\omega' - \omega)t}}{\omega - \omega'}$$

$$= -\frac{1 - \cos\big((\omega' - \omega)t\big)}{\omega' - \omega} - i \frac{\sin\big((\omega' - \omega)t\big)}{\omega' - \omega} \, .$$

$$(2.5.35)$$

The real part

$$\frac{1 - \cos\big((\omega' - \omega)t\big)}{\omega' - \omega} \cong \begin{cases} \dfrac{1}{\omega' - \omega} & \text{for} \quad \text{large } t \, , \\ 0 & \text{for} \quad t = 0 \, , \end{cases} \qquad (2.5.36)$$

where "large t" means large on the scale set by typical $\omega' - \omega$ differences, has the characterizing properties of the principal value of a singular integration,

$$\frac{1 - \cos\big((\omega' - \omega)t\big)}{\omega' - \omega} \cong \mathcal{P} \frac{1}{\omega' - \omega} \, . \qquad (2.5.37)$$

The imaginary part

$$\frac{\sin\big((\omega' - \omega)t\big)}{\omega' - \omega} \cong \begin{cases} t & \text{for} \quad \omega' = \omega \, , \\ 0 & \text{for} \quad \omega' \neq \omega \, , \end{cases} \qquad (2.5.38)$$

has the features characteristic of a δ function,

$$\frac{\sin\big((\omega' - \omega)t\big)}{\omega' - \omega} \cong \pi\delta(\omega' - \omega) \qquad (2.5.39)$$

for large t. Together we have

$$
\begin{aligned}
\frac{1}{2}\Gamma &= i \int d\omega' \, f(\omega') \left[-\mathcal{P} \frac{1}{\omega' - \omega} - i\pi\delta(\omega' - \omega) \right] \\
&= \pi f(\omega) - i\mathcal{P} \int d\omega' \, \frac{f(\omega')}{\omega' - \omega} \\
&= \frac{1}{2}\gamma + i\Delta\omega
\end{aligned}
\tag{2.5.40}
$$

with the transition rate $\gamma = 2\pi f(\omega)$ of (2.5.15) that we found in Section 2.5.1 as an application of the golden rule. In addition, we now identify a frequency shift

$$
\Delta\omega = -\mathcal{P} \int d\omega' \, \frac{f(\omega')}{\omega' - \omega}
\tag{2.5.41}
$$

which is very tiny and only noticeable in high-precision spectroscopy. It is part of the so-called *Lamb shift*, a very fine detail in atomic spectra, discovered by Willis E. Lamb.

In summary, we have

$$
\alpha(t) \cong e^{-i(\omega + \Delta\omega)t} e^{-\gamma t/2}
\tag{2.5.42}
$$

and

$$
|\alpha(t)|^2 \cong e^{-\gamma t}
\tag{2.5.43}
$$

is the probability that no photon emission has occurred up to time t. To get a better feeling for what is going on, let us look at

$$
\begin{aligned}
\beta_\nu(t) &= i\Omega_\nu \int_0^t dt' \, e^{-i\omega_\nu(t - t')} \alpha(t') \\
&\cong i\Omega_\nu e^{-i\omega_\nu t} \int_0^t dt' \, e^{i\omega_\nu t'} e^{-i(\omega + \Delta\omega)t'} e^{-\gamma t'/2} \\
&= i\Omega_\nu e^{-i\omega_\nu t} \frac{e^{i\omega_\nu t} e^{-i(\omega + \Delta\omega)t} e^{-\gamma t/2} - 1}{i(\omega_\nu - \omega - \Delta\omega) - \gamma/2} \\
&= e^{-i\omega_\nu t} \Omega_\nu \frac{e^{-i(\omega + \Delta\omega - \omega_\nu)t} e^{-\gamma t/2} - 1}{\omega_\nu - \omega - \Delta\omega + i\gamma/2}
\end{aligned}
\tag{2.5.44}
$$

which, for $t \gg 1/\gamma$, gives

$$
|\beta_\nu|^2 = |\Omega_\nu|^2 \frac{1}{(\omega_\nu - \omega - \Delta\omega)^2 + (\gamma/2)^2} .
\tag{2.5.45}
$$

The total emission probability, after waiting long enough, is therefore

$$\sum_{\nu} |\beta_{\nu}(t)|^2 \Big|_{t \gg 1/\gamma} = \sum_{\nu} |\Omega_{\nu}|^2 \frac{1}{(\omega_{\nu} - \omega - \Delta\omega)^2 + (\gamma/2)^2}$$

$$= \int d\omega' \, f(\omega') \frac{1}{(\omega' - \omega - \Delta\omega)^2 + (\gamma/2)^2}$$

$$= \int d\omega' \, \frac{\gamma}{2\pi} \frac{1}{(\omega' - \omega - \Delta\omega)^2 + (\gamma/2)^2}$$

$$= 1, \qquad (2.5.46)$$

where we first recognize that $\dfrac{1}{(\omega' - \omega - \Delta\omega)^2 + (\gamma/2)^2}$ is very strongly peaked at $\omega' = \omega + \Delta\omega \cong \omega$ with a narrow width given by γ:

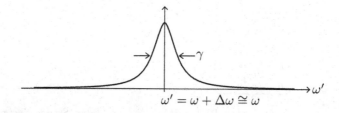

$$\omega' = \omega + \Delta\omega \cong \omega$$

and then note that, therefore, $f(\omega') \cong f(\omega + \Delta\omega) \cong f(\omega) = \gamma/(2\pi)$ is a permissible and consistent approximation.

This curve is the so-called *Lorentz profile*, named after Hendrik A. Lorentz, which is typical for the line shape of a spectral line, with its "full width at half maximum" equal to the transition rate γ. Under ideal experimental circumstances (no line broadening by thermal motion and the like), one observes this line shape in spectroscopy for well-isolated spectral lines.

2.5.3 *An exact treatment*

An alternative, and perhaps more systematic, way of solving the integral-differential equation (2.5.28) begins with incorporating the mode averaging of (2.5.33) right away, thereby arriving at

$$\frac{\partial}{\partial t} \left(e^{i\omega t} \alpha(t) \right) = -\int d\omega' \, f(\omega') \int_0^t dt' \, e^{i(\omega - \omega')(t - t')} \left(e^{i\omega t'} \alpha(t') \right). \qquad (2.5.47)$$

Now, rather than searching for $e^{i\omega t}\alpha(t)$ directly, we look for its Laplace transform (Marquis de Pierre S. Laplace),

$$a(s) = \int_0^\infty dt\, e^{-st}\, e^{i\omega t}\alpha(t)\,, \tag{2.5.48}$$

for which purpose we multiply (2.5.47) by e^{-st} and integrate over t. On the left-hand side this gives

$$\int_0^\infty dt\, e^{-st}\frac{\partial}{\partial t}\left(e^{i\omega t}\alpha(t)\right)$$

$$= e^{-st}\, e^{i\omega t}\alpha(t)\Big|_{t=0}^{t=\infty} + s\int_0^\infty dt\, e^{-st}\, e^{i\omega t}\alpha(t)$$

$$= -1 + sa(s)\,. \tag{2.5.49}$$

On the right-hand side we encounter

$$\underbrace{\int_0^\infty dt \int_0^t dt'}\ e^{-st}\, e^{i(\omega-\omega')(t-t')}\, e^{i\omega t'}\alpha(t')$$

$$= \int_0^\infty dt' \int_{t'}^\infty dt \qquad [\text{cf. (3.3.6) in } \textit{Simple Systems}]$$

$$= \int_0^\infty dt' \int_0^\infty dt\, e^{-s(t+t')}\, e^{i(\omega-\omega')t}\, e^{i\omega t'}\alpha(t')$$

$$= \int_0^\infty dt\, e^{-st}\, e^{i(\omega-\omega')t} \int_0^\infty dt'\, e^{-st'}\, e^{i\omega t'}\alpha(t')$$

$$= \frac{1}{s + i(\omega'-\omega)}a(s)\,, \tag{2.5.50}$$

where we observe the usual factorization of the Laplace transform of a convolution integral. This turns (2.5.47) into

$$sa(s) - 1 = -a(s)\int d\omega'\, \frac{f(\omega')}{s + i(\omega'-\omega)}\,. \tag{2.5.51}$$

Its immediate implication

$$a(s) = \left[s + \int d\omega'\, \frac{f(\omega')}{s + i(\omega'-\omega)}\right]^{-1} \tag{2.5.52}$$

yields the exact solution of (2.5.47) by virtue of the inverse Laplace transform that expresses $e^{i\omega t}\alpha(t)$ in terms of $a(s)$,

$$e^{i\omega t}\alpha(t) = \int_{-\infty}^{\infty} \frac{d\kappa}{2\pi} e^{(s_0 + i\kappa)t} a(s_0 + i\kappa) \quad \text{with} \quad s_0 > 0, \quad (2.5.53)$$

where we integrate $a(s)$ in the complex s plane along a line parallel to the imaginary axis with a positive, but otherwise arbitrary, real part of s. With the explicit expression for $a(s)$ in (2.5.52), and taking the limit $\epsilon \equiv s_0 \to 0$, we have

$$e^{i\omega t}\alpha(t) = \int_{-\infty}^{\infty} \frac{d\kappa}{2\pi i} e^{i\kappa t} \left[\kappa - i\epsilon - \int d\omega' \frac{f(\omega')}{\omega' - \omega + \kappa - i\epsilon} \right]^{-1} \Bigg|_{0 < \epsilon \to 0},$$
$$(2.5.54)$$

which could now serve as the basis for systematic approximations.

This exact solution of (2.5.47) states $\alpha(t)$ for all $t > 0$, which is much more detailed information than we care about. For, we must not forget that the physical model that is defined by the Hamilton operator (2.5.8) involves *physical approximations*, in particular the reduction of the complex atom to just two relevant states. Therefore, there is no need for utter mathematical precision when solving the coupled Schrödinger equations of (2.5.21). Rather, we should be guided by the physical situation and employ reasonable *mathematical approximations* when extracting the physical quantities of interest from (2.5.52).

With the aim of establishing contact with the findings in Section 2.5.2, we focus on the behavior for large t, which corresponds to small s values. The simplest approximation is then to replace the integral in (2.5.52) by its $s = 0$ value,

$$\int d\omega' \frac{f(\omega')}{s + i(\omega' - \omega)} \Bigg|_{s \to 0} = \int d\omega' f(\omega') \left[\pi\delta(\omega' - \omega) - i\mathcal{P}\frac{1}{\omega' - \omega} \right]$$
$$= \pi f(\omega) - i\mathcal{P} \int d\omega' \frac{f(\omega')}{i(\omega' - \omega)}$$
$$= \frac{1}{2}\gamma + i\Delta\omega, \quad (2.5.55)$$

where we recognize the ingredients of (2.5.40), the transition rate γ of (2.5.15) and the frequency shift $\Delta\omega$ of (2.5.41). The first step in (2.5.55) is an application of

$$\left.\frac{1}{x \pm i\epsilon}\right|_{\epsilon \to 0} = \mathcal{P}\frac{1}{x} \mp i\pi\delta(x), \qquad (2.5.56)$$

which combines the familiar models $\dfrac{x}{x^2 + \epsilon^2}$ and $\dfrac{\epsilon/\pi}{x^2 + \epsilon^2}$ for the principal value and the Dirac δ function, respectively.

With this simplest approximation then, we have

$$a(s) \cong \frac{1}{s + \frac{1}{2}\Gamma} = \int_0^\infty dt\, e^{-st}e^{-\frac{1}{2}\Gamma t} \qquad (2.5.57)$$

and are thus led back to the *ansatz* in (2.5.29). The Laplace transform reasoning, therefore, justifies this *ansatz* and offers the option of systematic improvements on the basis of (2.5.52).

2.6   Driven two-level atom

2.6.1   *Schrödinger equation*

In the situation considered in Section 2.5, there are no photons initially. The other extreme is the situation in which there are so many photons initially that we can regard the photon field as a classical radiation field. Then the emission of one more photon, or the absorption of one of the many photons, does not change the photon field significantly, and we can therefore neglect the back action of the atom on the photons. Formally, you can think of the photon state to be a collection of coherent states (see Section 3.4.2 in *Simple Systems*), that is: eigenkets of all A_ν, eigenbras of all $A_\nu^\dagger$, so that

(1) the initial state $\big|(e \text{ or } g), \{a_\nu'\}, t_2\big\rangle = \big|\text{init}\big\rangle$
     is a joint eigenket of all $A_\nu(t_2)$:

$$A_\nu(t_2)\big|\text{init}\big\rangle = \big|\text{init}\big\rangle a_\nu'\,; \qquad (2.6.1)$$

(2) the final state $\big\langle(e \text{ or } g), \{a_\nu^*\}, t_1\big| = \big\langle\text{fin}\big|$
     is a joint eigenbra of all $A_\nu(t_1)^\dagger$:

$$\big\langle\text{fin}\big|A_\nu^\dagger(t_1) = a_\nu^*\big\langle\text{fin}\big|\,. \qquad (2.6.2)$$

With no action of the atom on the photons, we have

$$A_\nu(t) = e^{-i\omega_\nu(t-t_2)} A_\nu(t_2) \,,$$
$$A_\nu(t)^\dagger = A_\nu(t_1)^\dagger \, e^{-i\omega_\nu(t_1-t)} \,,$$
$$\text{and} \quad a_\nu^* = \left(e^{-i\omega_\nu(t_1-t_2)} a_\nu' \right)^* = a_\nu'^{\,*} \, e^{i\omega_\nu(t_1-t_2)} \,. \tag{2.6.3}$$

Effectively, then, the operators $A_\nu(t), A_\nu^\dagger(t)$ are replaced by the numbers

$$A_\nu(t) \rightarrow e^{-i\omega_\nu(t-t_2)} a_\nu' \,,$$
$$A_\nu(t)^\dagger \rightarrow a_\nu^* \, e^{-i\omega_\nu(t_1-t)} = a_\nu'^{\,*} \, e^{i\omega_\nu(t-t_2)} \,, \tag{2.6.4}$$

and so H_{phot} of (2.5.4) is replaced by

$$H_{\text{phot}} \rightarrow \sum_\nu \hbar\omega_\nu |a_\nu'|^2 \,, \tag{2.6.5}$$

which is a number, and H_{int} is replaced by

$$H_{\text{int}} \rightarrow -\sum_\nu \hbar \left[\Omega_\nu \sigma a_\nu'^{\,*} \, e^{i\omega_\nu(t-t_2)} + \Omega_\nu^* \sigma^\dagger a_\nu' \, e^{-i\omega_\nu(t-t_2)} \right]$$
$$= -\hbar \left[\Omega(t)\sigma + \Omega(t)^* \sigma^\dagger \right] \tag{2.6.6}$$

with the numerical function

$$\Omega(t) = \sum_\nu \Omega_\nu a_\nu'^{\,*} \, e^{i\omega_\nu(t-t_2)} \,, \tag{2.6.7}$$

which is a rather arbitrary function of time.

In a typical experimental situation, the atom would be exposed to a laser pulse and the *time-dependent Rabi frequency* $\Omega(t)$ would specify the strength, duration, and spectral content of the pulse. In short, $\Omega(t)$ is under the control of the experimenter.

We have no further use for H_{phot}, now that it is just a numerical constant, so that

$$H = \hbar\omega\sigma^\dagger\sigma - \hbar\left(\Omega(t)\sigma + \Omega(t)^*\sigma^\dagger\right) \tag{2.6.8}$$

is the Hamilton operator for such a driven two-level atom. We take the general ket

$$| \, \rangle = |e,t\rangle\alpha(t) + |g,t\rangle\beta(t) \tag{2.6.9}$$

and write down the Schrödinger equations obeyed by the probability amplitudes

$$\alpha(t) = \langle e, t| \ \rangle \quad \text{and} \quad \beta(t) = \langle g, t| \ \rangle. \tag{2.6.10}$$

They are

$$\begin{aligned}
\mathrm{i}\hbar\frac{\partial}{\partial t}\alpha(t) &= \langle e, t|H| \ \rangle \\
&= \hbar\omega\langle e, t| \ \rangle - \hbar\Omega(t)^*\langle g, t| \ \rangle \\
&= \hbar\omega\alpha(t) - \hbar\Omega(t)^*\beta(t)
\end{aligned} \tag{2.6.11}$$

and

$$\begin{aligned}
\mathrm{i}\hbar\frac{\partial}{\partial t}\beta(t) &= \langle g, t|H| \ \rangle = -\hbar\Omega(t)\langle e, t| \ \rangle \\
&= -\hbar\Omega(t)\alpha(t),
\end{aligned} \tag{2.6.12}$$

or

$$\begin{aligned}
\left(\frac{\partial}{\partial t} + \mathrm{i}\omega\right)\alpha(t) &= \mathrm{i}\Omega(t)^*\beta(t), \\
\frac{\partial}{\partial t}\beta(t) &= \mathrm{i}\Omega(t)\alpha(t).
\end{aligned} \tag{2.6.13}$$

Upon recognizing once more that

$$\left(\frac{\partial}{\partial t} + \mathrm{i}\omega\right)\alpha(t) = \mathrm{e}^{-\mathrm{i}\omega t}\frac{\partial}{\partial t}\left(\mathrm{e}^{\mathrm{i}\omega t}\alpha(t)\right), \tag{2.6.14}$$

this pair of equations can be presented quite compactly as

$$\frac{\partial}{\partial t}\begin{pmatrix} \mathrm{e}^{\mathrm{i}\omega t}\alpha(t) \\ \beta(t) \end{pmatrix} = \mathrm{i}\begin{pmatrix} 0 & \mathrm{e}^{\mathrm{i}\omega t}\Omega(t)^* \\ \mathrm{e}^{-\mathrm{i}\omega t}\Omega(t) & 0 \end{pmatrix}\underbrace{\begin{pmatrix} \mathrm{e}^{\mathrm{i}\omega t}\alpha(t) \\ \beta(t) \end{pmatrix}}_{=\psi(t)} \tag{2.6.15}$$

or

$$\frac{\partial}{\partial t}\psi(t) = \mathrm{i}\begin{pmatrix} 0 & \widetilde{\Omega}(t)^* \\ \widetilde{\Omega}(t) & 0 \end{pmatrix}\psi(t) \tag{2.6.16}$$

where $\psi(t)$ is the two-component column composed of the probability amplitudes $\mathrm{e}^{\mathrm{i}\omega t}\alpha(t)$ and $\beta(t)$, and

$$\widetilde{\Omega}(t) = \mathrm{e}^{-\mathrm{i}\omega t}\Omega(t). \tag{2.6.17}$$

The column $\psi(t)$ is the numerical description of the state ket of (2.6.9) and, accordingly, (2.6.16) is the Schrödinger equation for the driven two-level atom.

Inasmuch as the replacements $\alpha(t) \to e^{i\omega t}\alpha(t)$, $\Omega \to \widetilde{\Omega}$ remove the time dependence that is there without the interaction, so that $\psi(t)$ changes only as a consequence of the interaction, we have arrived at the equation of motion in the "interaction picture". Never mind the words, though, what is important is that, even after this convenient rewriting, we cannot solve the equation for an arbitrary time dependence $\widetilde{\Omega}(t)$. The reason is, of course, that the 2×2 matrices for different times do not commute and, therefore, we cannot integrate the differential equation by a simple exponentiation.

2.6.2 Resonant drive

There is, however, an important exception namely that of *resonant driving*, meaning that the frequency of the external drive exactly matches the natural frequency of the atomic transition. Here this means

$$\Omega(t) = \underset{\underset{\text{constant}}{\uparrow}}{\Omega}\, e^{i\omega t}\,, \qquad (2.6.18)$$

so that

$$\widetilde{\Omega}(t) = \Omega = \text{const}\,, \qquad (2.6.19)$$

and the parametric time dependence disappears from the differential equation for $\psi(t)$,

$$\frac{\partial}{\partial t}\psi(t) = i\begin{pmatrix} 0 & \Omega^* \\ \Omega & 0 \end{pmatrix}\psi(t)\,. \qquad (2.6.20)$$

This 2×2 matrix is, essentially, a square root of the identity, inasmuch as

$$\begin{pmatrix} 0 & \Omega^* \\ \Omega & 0 \end{pmatrix}^2 = |\Omega|^2\begin{pmatrix} 1 & 0 \\ 0 & 1 \end{pmatrix}\,. \qquad (2.6.21)$$

As a consequence, it is easily exponentiated,

$$e^{i\begin{pmatrix} 0 & \Omega^* \\ \Omega & 0 \end{pmatrix}t} = \cos\left(\begin{pmatrix} 0 & \Omega^* \\ \Omega & 0 \end{pmatrix}t\right) + i\sin\left(\begin{pmatrix} 0 & \Omega^* \\ \Omega & 0 \end{pmatrix}t\right)$$

$$= \cos(|\Omega|t)\begin{pmatrix} 1 & 0 \\ 0 & 1 \end{pmatrix} + i\sin(|\Omega|t)\frac{1}{|\Omega|}\begin{pmatrix} 0 & \Omega^* \\ \Omega & 0 \end{pmatrix}\,, \qquad (2.6.22)$$

and

$$\psi(t) = e^{i\begin{pmatrix} 0 & \Omega^* \\ \Omega & 0 \end{pmatrix}t}\psi(0) \qquad (2.6.23)$$

reads, more explicitly,

$$e^{i\omega t}\alpha(t) = \cos(|\Omega|t)\alpha(0) + i\sin(|\Omega|t)\frac{\Omega^*}{|\Omega|}\beta(0)\,,$$

$$\beta(t) = \cos(|\Omega|t)\beta(0) + i\sin(|\Omega|t)\frac{\Omega}{|\Omega|}\alpha(0)\,. \qquad (2.6.24)$$

These probability amplitudes, and then also the resulting probabilities, are periodic functions in time with the period given by the single frequency that is still present, the *Rabi frequency* $|\Omega|$.

2.6.3   *Periodic drive*

What about a *periodic drive* at a frequency different from ω? Let us consider

$$\Omega(t) = \Omega\,e^{i(\omega + \Delta)t}\,, \qquad (2.6.25)$$

where the *detuning* Δ is the frequency mismatch, so that

$$\widetilde{\Omega}(t) = \Omega\,e^{i\Delta t} \quad \text{with} \quad \Omega = \text{const.} \qquad (2.6.26)$$

Then

$$\begin{pmatrix} 0 & \widetilde{\Omega}(t)^* \\ \widetilde{\Omega}(t) & 0 \end{pmatrix} = \begin{pmatrix} 0 & \Omega^*\,e^{-i\Delta t} \\ \Omega\,e^{i\Delta t} & 0 \end{pmatrix}$$

$$= \begin{pmatrix} e^{-i\Delta t/2} & 0 \\ 0 & e^{i\Delta t/2} \end{pmatrix}\begin{pmatrix} 0 & \Omega^* \\ \Omega & 0 \end{pmatrix}\begin{pmatrix} e^{i\Delta t/2} & 0 \\ 0 & e^{-i\Delta t/2} \end{pmatrix}$$

$$(2.6.27)$$

is an invitation to take a look at

$$\frac{\partial}{\partial t}\left[\begin{pmatrix} e^{i\Delta t/2} & 0 \\ 0 & e^{-i\Delta t/2} \end{pmatrix}\psi\right] = i\begin{pmatrix} 0 & \Omega^* \\ \Omega & 0 \end{pmatrix}\begin{pmatrix} e^{i\Delta t/2} & 0 \\ 0 & e^{-i\Delta t/2} \end{pmatrix}\psi$$

$$+ i\begin{pmatrix} \Delta/2 & 0 \\ 0 & -\Delta/2 \end{pmatrix}\begin{pmatrix} e^{i\Delta t/2} & 0 \\ 0 & e^{-i\Delta t/2} \end{pmatrix}\psi$$

$$(2.6.28)$$

so that

$$\frac{\partial}{\partial t}\overline{\psi}(t) = \mathrm{i}\begin{pmatrix} \Delta/2 & \Omega^* \\ \Omega & -\Delta/2 \end{pmatrix}\overline{\psi}(t) \qquad (2.6.29)$$

is a differential equation, with no parametric time dependence, for

$$\overline{\psi}(t) = \begin{pmatrix} \mathrm{e}^{\mathrm{i}\Delta t/2} & 0 \\ 0 & \mathrm{e}^{-\mathrm{i}\Delta t/2} \end{pmatrix}\psi(t).$$

It can, therefore, be integrated by exponentiation,

$$\overline{\psi}(t) = \mathrm{e}^{\mathrm{i}\begin{pmatrix} \Delta/2 & \Omega^* \\ \Omega & -\Delta/2 \end{pmatrix}t}\,\overline{\psi}(0). \qquad (2.6.30)$$

The 2×2 matrix here is also essentially a square root of the identity,

$$\begin{pmatrix} \Delta/2 & \Omega^* \\ \Omega & -\Delta/2 \end{pmatrix}^2 = \left(|\Omega|^2 + (\Delta/2)^2\right)\begin{pmatrix} 1 & 0 \\ 0 & 1 \end{pmatrix}, \qquad (2.6.31)$$

and so we get

$$\overline{\psi}(t) = \left[\cos(\overline{\Omega}t)\begin{pmatrix} 1 & 0 \\ 0 & 1 \end{pmatrix} + \mathrm{i}\frac{\sin(\overline{\Omega}t)}{\overline{\Omega}}\begin{pmatrix} \Delta/2 & \Omega^* \\ \Omega & -\Delta/2 \end{pmatrix}\right]\overline{\psi}(0) \qquad (2.6.32)$$

with

$$\overline{\Omega} = \sqrt{|\Omega|^2 + (\Delta/2)^2}. \qquad (2.6.33)$$

The oscillation is now with the frequency of this *modified Rabi frequency*, modified by the detuning Δ between the natural frequency ω of the atomic transition and the frequency $\omega + \Delta$ of the external driving field.

2-9 Suppose $\overline{\psi}(0) = \begin{pmatrix} 0 \\ 1 \end{pmatrix}$, so that the atom is deexcited at time $t = 0$. What is the probability of having the atom excited at any later time t? At which times is the excitation probability largest? How large is it then?

2-10 What are the eigenvalues and eigencolumns of $\begin{pmatrix} \Delta/2 & \Omega^* \\ \Omega & -\Delta/2 \end{pmatrix}$? What is their physical significance? Hint: You may find the parameterization $\Delta/2 = \overline{\Omega}\cos(2\vartheta)$, $\Omega = \overline{\Omega}\,\mathrm{e}^{\mathrm{i}\varphi}\sin(2\vartheta)$ useful.

2.6.4 Very slow drive. Adiabatic evolution

There is one more situation in which we can say something quite definite about the evolution governed by the Hamilton operator (2.6.8),

$$H = \hbar\omega\sigma^\dagger\sigma - \hbar\big(\Omega(t)\sigma + \Omega(t)^*\sigma^\dagger\big), \qquad (2.6.34)$$

namely when the time-dependent Rabi frequency $\Omega(t)$ changes very slowly — that is: slowly on the time scale set by the energy difference between the eigenstates of H_0. These eigenstates and their eigenvalues themselves depend parametrically on time t, because $\Omega(t)$ introduces such a time dependence into the Hamilton operator. We summarize the Schrödinger equations (2.6.13) for $\alpha(t) = \langle e, t | \ \rangle$, $\beta(t) = \langle g, t | \ \rangle$ in

$$i\hbar\frac{\partial}{\partial t}\begin{pmatrix}\alpha(t)\\\beta(t)\end{pmatrix} = \hbar\begin{pmatrix}\omega & -\Omega(t)^*\\-\Omega(t) & 0\end{pmatrix}\begin{pmatrix}\alpha(t)\\\beta(t)\end{pmatrix}$$

$$\equiv \mathcal{H}(t)\begin{pmatrix}\alpha(t)\\\beta(t)\end{pmatrix} \qquad (2.6.35)$$

where the 2×2 matrix $\mathcal{H}(t)$ is the numerical matrix representation for H as referring to the basis formed by $|e, t\rangle$ and $|g, t\rangle$,

$$H = \big(|e, t\rangle, |g, t\rangle\big)\mathcal{H}\begin{pmatrix}\langle e, t|\\\langle g, t|\end{pmatrix}. \qquad (2.6.36)$$

We find the eigenvalues and eigenstates of H by determining the eigenvalues, eigencolumns, and eigenrows of matrix $\mathcal{H}$. To this end, we note that the trace of $\mathcal{H}$ is $\hbar\omega$ and the determinant is $-\hbar^2|\Omega(t)|^2$. It follows that the eigenvalues are given by

$$E_\pm = \frac{1}{2}\hbar\omega \pm \frac{1}{2}\hbar\sqrt{\omega^2 + 4|\Omega(t)|^2}. \qquad (2.6.37)$$

One verifies easily that $E_+ + E_- = \hbar\omega$ is the correct value of the trace and $E_+E_- = -\hbar^2|\Omega(t)|^2$ is the correct value of the determinant.

It will be expedient to write

$$\overline{\omega} = \sqrt{\omega^2 + 4|\Omega(t)|^2}\,,$$
$$\omega = \overline{\omega}\cos(2\vartheta)\,,$$
$$\Omega = \frac{1}{2}\overline{\omega}\,e^{i\varphi}\sin(2\vartheta)\,, \qquad (2.6.38)$$

where $\overline{\omega}, \vartheta$, and φ all depend on t as they inherit the parametric t dependence of $\Omega(t)$. Then,

$$
\mathcal{H} = \hbar \begin{pmatrix} \omega & -\Omega^* \\ -\Omega & 0 \end{pmatrix} = \frac{1}{2}\hbar\omega \begin{pmatrix} 1 & 0 \\ 0 & 1 \end{pmatrix} + \frac{\hbar}{2} \begin{pmatrix} \omega & -2\Omega^* \\ -2\Omega & -\omega \end{pmatrix}
$$

$$
= \frac{1}{2}\hbar\omega \begin{pmatrix} 1 & 0 \\ 0 & 1 \end{pmatrix} + \frac{1}{2}\hbar\overline{\omega} \begin{pmatrix} \cos(2\vartheta) & -\mathrm{e}^{-\mathrm{i}\varphi}\sin(2\vartheta) \\ -\mathrm{e}^{\mathrm{i}\varphi}\sin(2\vartheta) & -\cos(2\vartheta) \end{pmatrix}, \quad (2.6.39)
$$

and since

$$
\begin{pmatrix} \cos(2\vartheta) & -\mathrm{e}^{-\mathrm{i}\varphi}\sin(2\vartheta) \\ -\mathrm{e}^{\mathrm{i}\varphi}\sin(2\vartheta) & -\cos(2\vartheta) \end{pmatrix}
$$

$$
= \begin{pmatrix} \cos\vartheta \\ -\mathrm{e}^{\mathrm{i}\varphi}\sin\vartheta \end{pmatrix} \left(\cos\vartheta, \, -\mathrm{e}^{-\mathrm{i}\varphi}\sin\vartheta \right) - \begin{pmatrix} \mathrm{e}^{-\mathrm{i}\varphi}\sin\vartheta \\ \cos\vartheta \end{pmatrix} \left(\mathrm{e}^{\mathrm{i}\varphi}\sin\vartheta, \, \cos\vartheta \right)
$$

$$
(2.6.40)
$$

exhibits the eigencolumns and eigenrows of this 2×2 matrix to its eigenvalues $+1$ and -1, we have, for $\mathcal{H}$ itself,

$$
\text{eigenvalue } \frac{\hbar}{2}(\omega + \overline{\omega}) : \quad \text{eigencolumn } \begin{pmatrix} \cos\vartheta \\ -\mathrm{e}^{\mathrm{i}\varphi}\sin\vartheta \end{pmatrix},
$$

$$
\text{eigenrow } \left(\cos\vartheta, \, -\mathrm{e}^{-\mathrm{i}\varphi}\sin\vartheta \right);
$$

$$
\text{eigenvalue } \frac{\hbar}{2}(\omega - \overline{\omega}) : \quad \text{eigencolumn } \begin{pmatrix} \mathrm{e}^{-\mathrm{i}\varphi}\sin\vartheta \\ \cos\vartheta \end{pmatrix},
$$

$$
\text{eigenrow } \left(\mathrm{e}^{\mathrm{i}\varphi}\sin\vartheta, \, \cos\vartheta \right). \quad (2.6.41)
$$

Suppose now that $\Omega(t)$ changes *slowly* in time, meaning roughly that

$$
\underbrace{\left| \frac{\partial \Omega}{\partial t} \middle/ \Omega \right|}_{\substack{\text{relative change} \\ \text{of } \Omega \text{ per unit time}}} \ll \underbrace{\frac{E_+ - E_-}{\hbar} = \overline{\omega}}_{\substack{\uparrow \\ 2\pi \text{ times the number of} \\ \text{oscillations per unit time}}}, \quad (2.6.42)
$$

then we expect that the system adapts itself to the slowly changing circumstances *adiabatically*. More specifically, we expect that if we are in the ground state of H initially, we stay in the ground state during the slow change of $\Omega(0) = 0$ to $\Omega(\text{late } t) = \Omega_{\text{fin}}$:

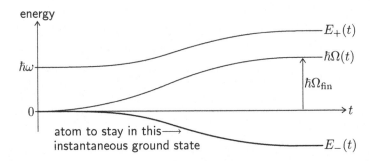

To see whether this expectation is correct, we look at the time dependence of

the probability amplitude for the instantaneous excited state:
$$a(t) = \left(\cos \vartheta , \, -\mathrm{e}^{-\mathrm{i}\varphi} \sin \vartheta \right) \begin{pmatrix} \alpha(t) \\ \beta(t) \end{pmatrix} ,$$
the probability amplitude for the instantaneous ground state:
$$b(t) = \left(\, \mathrm{e}^{\mathrm{i}\varphi} \sin \vartheta , \, \cos \vartheta \right) \begin{pmatrix} \alpha(t) \\ \beta(t) \end{pmatrix} . \qquad (2.6.43)$$

For these we have

$$\mathrm{i}\hbar \frac{\partial}{\partial t} a(t) = \left[\mathrm{i}\hbar \frac{\partial}{\partial t} \left(\cos \vartheta , \, -\mathrm{e}^{-\mathrm{i}\varphi} \sin \vartheta \right) \right] \begin{pmatrix} \alpha(t) \\ \beta(t) \end{pmatrix}$$
$$+ \left(\cos \vartheta , \, -\mathrm{e}^{-\mathrm{i}\varphi} \sin \vartheta \right) \mathcal{H} \begin{pmatrix} \alpha(t) \\ \beta(t) \end{pmatrix} \qquad (2.6.44)$$

and

$$\mathrm{i}\hbar \frac{\partial}{\partial t} b(t) = \left[\mathrm{i}\hbar \frac{\partial}{\partial t} \left(\, \mathrm{e}^{\mathrm{i}\varphi} \sin \vartheta , \, \cos \vartheta \right) \right] \begin{pmatrix} \alpha(t) \\ \beta(t) \end{pmatrix}$$
$$+ \left(\, \mathrm{e}^{\mathrm{i}\varphi} \sin \vartheta , \, \cos \vartheta \right) \mathcal{H} \begin{pmatrix} \alpha(t) \\ \beta(t) \end{pmatrix} , \qquad (2.6.45)$$

where the respective second terms have the eigenrows of $\mathcal{H}$ to the left of $\mathcal{H}$,
so that $\mathcal{H} \to \frac{1}{2}\hbar(\omega + \overline{\omega})$ in (2.6.44) and $\mathcal{H} \to \frac{1}{2}\hbar(\omega - \overline{\omega})$ in (2.6.45). Thus

we get

$$i\frac{\partial}{\partial t}a(t) = \frac{\omega + \overline{\omega}}{2}a(t) + \left(\text{terms proportional to } \frac{\partial \vartheta}{\partial t} \text{ and } \frac{\partial \varphi}{\partial t}\right),$$

$$i\frac{\partial}{\partial t}b(t) = \frac{\omega - \overline{\omega}}{2}b(t) + \left(\text{terms proportional to } \frac{\partial \vartheta}{\partial t} \text{ and } \frac{\partial \varphi}{\partial t}\right),$$

$$(2.6.46)$$

where the (terms ...) are small if $\Omega(t)$ changes slowly. These small terms couple $a(t)$ to $b(t)$, as you will see when you work out the following exercise.

2-11 Find the explicit form for the terms proportional to $\dfrac{\partial \vartheta}{\partial t}$ and $\dfrac{\partial \varphi}{\partial t}$.

But it is clear that they are negligible if they are indeed small on the scale set by the rapid evolution that happens with frequency $(E_+ - E_-)/\hbar = \overline{\omega}$, and so we discard them.

In this adiabatic approximation, then, the equations are solved immediately by

$$a(t) \cong a(0)\,e^{-\frac{i}{2}\int_0^t dt'\,(\omega + \overline{\omega}(t'))} \tag{2.6.47}$$

and

$$b(t) \cong b(0)\,e^{-\frac{i}{2}\int_0^t dt'\,(\omega - \overline{\omega}(t'))}, \tag{2.6.48}$$

so that

$$|a(t)|^2 \cong |a(0)|^2 \tag{2.6.49}$$

and

$$|b(t)|^2 \cong |b(0)|^2. \tag{2.6.50}$$

Yes, when the external conditions change adiabatically (here: $\dfrac{\partial \Omega}{\partial t}$ is small in comparison with $\overline{\omega}\Omega$), then the system goes from an instantaneous eigenstate of H to an instantaneous eigenstate of H with no significant probability of transitions between them.

2-12 Consider the 2×2 Hamilton matrix

$$\mathcal{H}(t) = \hbar\omega \begin{pmatrix} \cos(2\phi(t)) & \sin(2\phi(t)) \\ \sin(2\phi(t)) & -\cos(2\phi(t)) \end{pmatrix} \quad \text{with} \quad \phi(t) = \frac{\pi t}{T},$$

where $T > 0$. Use the eigencolumns $\psi_{\pm}(t)$ of $\mathcal{H}(t)$ to the eigenvalues $\pm\hbar\omega$ to write $\psi(t) = \begin{pmatrix} \alpha(t) \\ \beta(t) \end{pmatrix} = a(t)\psi_{+}(t) + b(t)\psi_{-}(t)$ and find the matrix $\mathcal{M}$ in

$$\frac{\partial}{\partial t} \begin{pmatrix} a(t) \\ b(t) \end{pmatrix} = i\mathcal{M} \begin{pmatrix} a(t) \\ b(t) \end{pmatrix}.$$

Check: If you get it right, $\mathcal{M}$ does not depend on t.

2-13 Solve this equation to find $\psi(T)$ for $\psi(0) = \begin{pmatrix} 1 \\ 0 \end{pmatrix}$. What is the dominating T dependence of the probability $|\beta(T)|^2$ for $\omega T \ll 1$? And what is it for $\omega T \gg 1$?

2.7 Adiabatic population transfer

Here is an application of adiabatic evolution to a real-life physical problem. Some atoms have degenerate ground states (magnetic sublevels) that can couple to the same excited state with electromagnetic radiation of the same frequency but different polarization. Then, the experimenter can separately control the time-dependent Rabi frequencies $\Omega_1(t)$ and $\Omega_2(t)$ in a level scheme like this one:

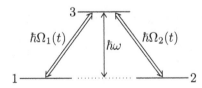

In quantum optics one speaks of a "three-level atom in Λ configuration". Ground states 1 and 2 are of energy 0 (by convention), excited state 3 has energy $\hbar\omega$; the coupling between 1 and 3 has a strength given by $\Omega_1(t)$, the coupling between 2 and 3 has a strength given by $\Omega_2(t)$. There is no direct coupling between 1 and 2. The 3×3 matrix for the Hamilton operator in this situation is therefore

$$\mathcal{H}(t) = \begin{pmatrix} \langle 1,t| \\ \langle 2,t| \\ \langle 3,t| \end{pmatrix} H\Big(|1,t\rangle, |2,t\rangle, |3,t\rangle\Big) = \hbar \begin{pmatrix} 0 & 0 & -\Omega_1^* \\ 0 & 0 & -\Omega_2^* \\ -\Omega_1 & -\Omega_2 & \omega \end{pmatrix} \qquad (2.7.1)$$

with $\Omega_1 = \Omega_1(t)$, $\Omega_2 = \Omega_2(t)$.

Now suppose that initially the atom is in ground state 1,

$$\psi_{\text{init}} = \begin{pmatrix} 1 \\ 0 \\ 0 \end{pmatrix}, \qquad (2.7.2)$$

and we want to choose $\Omega_1(t)$ and $\Omega_2(t)$ such that the atom is surely in ground state 2,

$$\psi_{\text{fin}} = \begin{pmatrix} 0 \\ 1 \\ 0 \end{pmatrix}, \qquad (2.7.3)$$

at the final time when $\Omega_1(t) = 0$ and $\Omega_2(t) = 0$ again. This can be accomplished indeed, and the crucial observation is that $\mathcal{H}$ has an eigencolumn to eigenvalue 0,

$$\mathcal{H}\begin{pmatrix} \Omega_2 \\ -\Omega_1 \\ 0 \end{pmatrix} = 0, \qquad (2.7.4)$$

for *any* choice of Ω_1 and Ω_2. In particular, we note

$$\Omega_1 = 0, \; \Omega_2 \neq 0 : \qquad \begin{pmatrix} \Omega_2 \\ -\Omega_1 \\ 0 \end{pmatrix} \propto \begin{pmatrix} 1 \\ 0 \\ 0 \end{pmatrix} = \psi_{\text{init}} \qquad (2.7.5)$$

and

$$\Omega_1 \neq 0, \; \Omega_2 = 0 : \qquad \begin{pmatrix} \Omega_2 \\ -\Omega_1 \\ 0 \end{pmatrix} \propto \begin{pmatrix} 0 \\ 1 \\ 0 \end{pmatrix} = \psi_{\text{fin}}. \qquad (2.7.6)$$

Therefore, if we arrange matters such that first $\Omega_2(t)$ acquires a sufficiently large value, while $\Omega_1(t) = 0$, and then $\Omega_1(t)$ increases, while $\Omega_2(t)$ decreases to $\Omega_2(t) = 0$, followed by the switching off of $\Omega_1(t)$, then the sequence

$$
\underbrace{\begin{pmatrix} 1 \\ 0 \\ 0 \end{pmatrix}}_{\substack{\text{initial} \\ \text{time}}} \longrightarrow \underbrace{\frac{1}{\sqrt{|\Omega_1|^2 + |\Omega_2|^2}} \begin{pmatrix} \Omega_2 \\ -\Omega_1 \\ 0 \end{pmatrix}}_{\substack{\text{intermediate} \\ \text{time}}} \longrightarrow \underbrace{\begin{pmatrix} 0 \\ 1 \\ 0 \end{pmatrix}}_{\substack{\text{final} \\ \text{time}}}
\tag{2.7.7}
$$

will be realized as an adiabatic evolution.

Graphically, we need to have

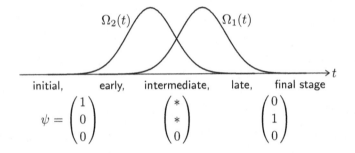

or, indicating level populations by circles ($\bullet = 100\%$, $\circ = $ part of 100%),

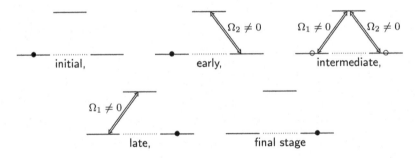

During the adiabatic transition from $\begin{pmatrix} 1 \\ 0 \\ 0 \end{pmatrix}$ to $\begin{pmatrix} 0 \\ 1 \\ 0 \end{pmatrix}$ there is never a component of $\begin{pmatrix} 0 \\ 0 \\ 1 \end{pmatrix}$, that is: during the whole process the atom will not be found in the excited level 3, which is very important in practice because from this excited level the atom can usually also make spontaneous transitions to other final states than the two ground states of interest.

2-14 What are the other two eigenvalues of the $\mathcal{H}(t)$ in (2.7.1) and what are the respective eigencolumns? Be sure to pay due attention to the proper normalization of the eigencolumns, and consider the limiting situations of $\Omega_1 \to 0$ while $\Omega_2 \neq 0$, $\Omega_2 \to 0$ while $\Omega_1 \neq 0$, as well as $\Omega_1 \to 0$ and $\Omega_2 \to 0$ simultaneously.

2.8   Equation of motion for the unitary evolution operator

In the situations discussed in Sections 2.6–2.7, the respective Hamilton operators possess parametric time dependences, so that the findings of Sections 2.1–2.3 do not apply immediately because they are all based on (2.1.1), where a parametric time dependence is explicitly excluded. This restriction can be lifted, however, as the formalism can be generalized quite easily and straightforwardly.

In relating bras at time t to those at the earlier time t_0,

$$\langle \dots, t| = \langle \dots, t_0 | U\big(A(t_0); t_0, t\big) , \qquad (2.8.1)$$

it is natural to regard the unitary evolution operator U as a function of the dynamical variables at time t_0, indicated by the argument $A(t_0)$. In addition it has, of course, a parametric dependence on t_0 and t. The dynamical variables at time t are related to those at time t_0 by the same unitary operator,

$$A(t) = U^\dagger A(t_0) U \quad \text{with} \quad U = U\big(A(t_0); t_0, t\big) . \qquad (2.8.2)$$

In particular, it follows then that

$$U\big(A(t); t_0, t\big) = \underbrace{U\big(A(t_0); t_0, t\big)^\dagger U\big(A(t_0); t_0, t\big)}_{=1} U\big(A(t_0); t_0, t\big)$$

$$= U\big(A(t_0); t_0, t\big) , \qquad (2.8.3)$$

that is: in the evolution operator U it makes no difference if we take the dynamical operators at the final or initial time, the functional dependence on them is exactly the same.

The Schrödinger equation

$$i\hbar \frac{\partial}{\partial t} \langle \dots, t| = \langle \dots, t| H\big(A(t), t\big) \qquad (2.8.4)$$

involves the Hamilton operator at time t, with the dynamical variables referring to time t as well. We compare with

$$i\hbar\frac{\partial}{\partial t}\langle\ldots,t| = \langle\ldots,t_0|i\hbar\frac{\partial}{\partial t}U(A(t_0);t_0,t)$$

$$= \langle\ldots,t|U(A(t_0);t_0,t)^{-1}i\hbar\frac{\partial}{\partial t}U(A(t_0);t_0,t) \quad (2.8.5)$$

and conclude that

$$i\hbar\frac{\partial}{\partial t}U(A(t_0);t_0,t) = U(A(t_0);t_0,t)H(A(t),t). \quad (2.8.6)$$

It is far more convenient to have all dynamical operators referring to the same instant, which we achieve by invoking

$$H(A(t),t) = U(A(t_0);t_0,t)^{\dagger}H(A(t_0),t)U(A(t_0);t_0,t) \quad (2.8.7)$$

to arrive at

$$i\hbar\frac{\partial}{\partial t}U(A(t_0);t_0,t) = H(A(t_0),t)U(A(t_0);t_0,t). \quad (2.8.8)$$

This is often referred to, somewhat misleadingly, as the *Schrödinger equation for U*. In it, we have the dynamical variables at the common time t_0 — and this is the important detail: the time argument is a common one. In fact, *any* common time will do, because the dynamical variables at one instant are related to those at another one by a unitary transformation and, equally important, the differentiation in this Schrödinger equation for U refers to the parametric t dependence only, not to the dynamical or total t dependence.

2-15 In particular, we can put $A(t_0) \to A(t)$ everywhere in (2.8.8). Do this, indeed, and compare $i\hbar\frac{\partial}{\partial t}U(A(t);t_0,t)$ with $i\hbar\frac{d}{dt}U(A(t);t_0,t)$.

Now, having agreed upon the common time that we wish to use for reference (either t_0 or t in most applications), we suppress the dependence on the dynamical variables and write, more compactly,

$$i\hbar\frac{\partial}{\partial t}U(t_0,t) = H(t)U(t_0,t). \quad (2.8.9)$$

The Hamilton operator is typically composed of an "unperturbed" part H_0 that usually has no parametric t dependence, and a perturbation $H_1(t)$ that usually does depend on t parametrically,

$$H(t) = H_0(t) + H_1(t). \quad (2.8.10)$$

As noted above, this is a bit more general than the situation in Sections 2.1–2.3 where $\frac{\partial}{\partial t}H = 0$ is assumed in (2.1.1).

We shall not insist on $\frac{\partial}{\partial t}H_0 = 0$, but allow a parametric time dependence in H_0 as well. For $H_1(t) \equiv 0$, we would then have

$$i\hbar\frac{\partial}{\partial t}U_0(t_0,t) = H_0(t)U_0(t_0,t) \tag{2.8.11}$$

with a *known* solution for $U_0(t_0,t)$. If $\frac{\partial}{\partial t}H_0 = 0$, indeed, we have the familiar

$$U_0(t_0,t) = e^{-\frac{i}{\hbar}(t-t_0)H_0}, \tag{2.8.12}$$

of course, which depends only on the duration $t - t_0$, not on t and t_0 individually.

We introduce the scattering operator $S(t_0,t)$ in analogy with what we did in (2.2.3) by writing

$$U(t_0,t) = U_0(t_0,t)S(t_0,t). \tag{2.8.13}$$

Then

$$i\hbar\frac{\partial}{\partial t}U(t_0,t) = \left[i\hbar\frac{\partial}{\partial t}U_0(t_0,t)\right]S(t_0,t) + U_0(t_0,t)i\hbar\frac{\partial}{\partial t}S(t_0,t)$$

$$= H_0(t)U_0(t_0,t)S(t_0,t) + U_0(t_0,t)i\hbar\frac{\partial}{\partial t}S(t_0,t), \tag{2.8.14}$$

and by the Schrödinger equation (2.8.8) for $U(t_0,t)$ also

$$i\hbar\frac{\partial}{\partial t}U(t_0,t) = \big(H_0(t) + H_1(t)\big)U(t_0,t)$$

$$= \big(H_0(t) + H_1(t)\big)U_0(t_0,t)S(t_0,t). \tag{2.8.15}$$

The implied equation of motion for the scattering operator is therefore

$$i\hbar\frac{\partial}{\partial t}S(t_0,t) = U_0(t_0,t)^{-1}H_1(t)U_0(t_0,t)S(t_0,t) \tag{2.8.16}$$

or

$$i\hbar\frac{\partial}{\partial t}S(t_0,t) = \overline{H_1}(t_0,t)S(t_0,t) \tag{2.8.17}$$

with

$$\overline{H_1}(t_0,t) = U_0(t_0,t)^{-1}H_1(t)U_0(t_0,t). \tag{2.8.18}$$

Upon incorporating the initial condition $U(t_0, t_0) = 1$ or $S(t_0, t_0) = 1$, we get

$$S(t_0, t) = 1 - \frac{i}{\hbar} \int_{t_0}^{t} dt' \, \overline{H_1}(t_0, t') S(t_0, t') \,, \qquad (2.8.19)$$

an equation of the Lippmann–Schwinger form. It is the obvious analog of the integral equation (2.2.10) and, owing to this analogy, we can immediately proceed to the Born series

$$S(t_0, t) = 1 - \frac{i}{\hbar} \int_{t_0}^{t} dt' \, \overline{H_1}(t_0, t)$$
$$+ \left(-\frac{i}{\hbar}\right)^2 \int_{t_0}^{t} dt' \, \overline{H_1}(t_0, t') \int_{t_0}^{t'} dt'' \, \overline{H_1}(t_0, t'')$$
$$+ \cdots \qquad (2.8.20)$$

and its formal summation in terms of Dyson's time-ordered exponential

$$S(t_0, t) = \left[\exp\left(-\frac{i}{\hbar} \int_{t_0}^{t} dt' \, \overline{H_1}(t_0, t') \right) \right]_{+} \,, \qquad (2.8.21)$$

where the time ordering refers to the second argument of $\overline{H_1}(t_0, t)$. There is also the iteration scheme of (2.2.11),

$$S_{n+1}(t_0, t) = 1 - \frac{i}{\hbar} \int_{t_0}^{t} dt' \, \overline{H_1}(t_0, t') S_n(t_0, t') \,, \qquad (2.8.22)$$

and if we start with $S(t_0, t) = 1$ this generates the successive approximations of the Born series.

2-16　Write

$$S(t_0, t) = \sum_{k=0}^{\infty} \frac{(-i/\hbar)^k}{k!} s_k(t_0, t)$$

where s_k involves k factors of $\overline{H_1}$. How do you get s_{k+1} from s_k?

2-17　Consider infinitesimal variations of $\overline{H_1}(t_0, t')$ at all intermediate times in (2.8.21) and establish the appropriate generalization of (1.4.41) to time-ordered exponentials.

In practice, the reliability and power of any such approximation method depends much on how one splits the Hamilton operator into H_0 and $H_1(t)$.

As an illustration, let us reconsider the situation of (2.6.25)–(2.6.33), that is a Hamilton operator represented by the 2×2 matrix

$$\mathcal{H} = \hbar \begin{pmatrix} \omega & -\Omega^* e^{-i\omega t} e^{-i\Delta t} \\ -\Omega e^{i\omega t} e^{i\Delta t} & 0 \end{pmatrix} \tag{2.8.23}$$

where Ω is constant in time. Here, the best choice is not the simple split

$$\mathcal{H} = \hbar \underbrace{\begin{pmatrix} \omega & 0 \\ 0 & 0 \end{pmatrix}}_{=\mathcal{H}_0} + \hbar \underbrace{\begin{pmatrix} 0 & -\Omega^* e^{-i\omega t} e^{-i\Delta t} \\ -\Omega e^{i\omega t} e^{i\Delta t} & 0 \end{pmatrix}}_{=\mathcal{H}_1(t)} \tag{2.8.24}$$

but rather the less obvious split

$$\mathcal{H} = \hbar \underbrace{\begin{pmatrix} \omega + \Delta/2 & 0 \\ 0 & -\Delta/2 \end{pmatrix}}_{=\mathcal{H}_0} + \hbar \underbrace{\begin{pmatrix} -\Delta/2 & -\Omega^* e^{-i\omega t} e^{-i\Delta t} \\ -\Omega e^{i\omega t} e^{i\Delta t} & \Delta/2 \end{pmatrix}}_{=\mathcal{H}_1(t)} \tag{2.8.25}$$

because then, for $t_0 = 0$,

$$U_0(t_0, t) \mathrel{\hat{=}} e^{-it\mathcal{H}_0/\hbar} = \begin{pmatrix} e^{-i\omega t} e^{-i\Delta t/2} & 0 \\ 0 & e^{i\Delta t/2} \end{pmatrix} \tag{2.8.26}$$

and

$$\begin{aligned}
\overline{H}_1(t) &\mathrel{\hat{=}} e^{it\mathcal{H}_0/\hbar} \mathcal{H}_1(t) e^{-it\mathcal{H}_0/\hbar} \\
&= \begin{pmatrix} e^{i\omega t} e^{i\Delta t/2} & 0 \\ 0 & e^{-i\Delta t/2} \end{pmatrix} \mathcal{H}_1(t) \begin{pmatrix} e^{-i\omega t} e^{-i\Delta t/2} & 0 \\ 0 & e^{i\Delta t/2} \end{pmatrix} \\
&= \hbar \begin{pmatrix} -\Delta/2 & -\Omega^* \\ -\Omega & \Delta/2 \end{pmatrix}
\end{aligned} \tag{2.8.27}$$

is independent of t. As a consequence the time-ordering becomes irrelevant and we have

$$\begin{aligned}
S(t_0 = 0, t) = e^{-it\overline{H}_1/\hbar} &\mathrel{\hat{=}} e^{it\begin{pmatrix} \Delta/2 & \Omega^* \\ \Omega & -\Delta/2 \end{pmatrix}} \\
&= \cos(\overline{\Omega} t) \begin{pmatrix} 1 & 0 \\ 0 & 1 \end{pmatrix} + i \frac{\sin(\overline{\Omega} t)}{\overline{\Omega}} \begin{pmatrix} \Delta/2 & \Omega^* \\ \Omega & -\Delta/2 \end{pmatrix}
\end{aligned} \tag{2.8.28}$$

with the modified Rabi frequency $\overline{\Omega} = \sqrt{|\Omega|^2 + (\Delta/2)^2}$ of (2.6.33).

Chapter 3

Scattering

3.1 Probability density, probability current density

If the physical system is described by the three-dimensional wave function

$$\psi(\vec{r}, t) = \langle \vec{r}, t | \ \rangle, \tag{3.1.1}$$

then the probability of finding it inside the volume V is

$$
\begin{aligned}
\text{prob(in } V, t) &= \int_V (\mathrm{d}\vec{r}) \, |\psi(\vec{r}, t)|^2 \\
&= \int_V (\mathrm{d}\vec{r}) \langle \ |\vec{r}, t\rangle\langle\vec{r}, t| \ \rangle
\end{aligned} \tag{3.1.2}
$$

which identifies

$$
\begin{aligned}
\rho(\vec{r}, t) &= \langle \ | \Big(|\vec{r}, t\rangle\langle\vec{r}, t| \Big) | \ \rangle \\
&= \Big\langle \delta\big(\vec{R}(t) - \vec{r}\big) \Big\rangle
\end{aligned} \tag{3.1.3}
$$

as the *probability density* associated with the actual state of the system. The latter version, that is: expectation value of $\delta\big(\vec{R}(t) - \vec{r}\big) = |\vec{r}, t\rangle\langle\vec{r}, t|$, carries over to mixed states for which a wave function is not available.

The probability prob(in V, t) changes in time because there is the possibility that $\rho(\vec{r}, t)$ increases or decreases inside V. This change comes about as the result of a *flux* of probability through the surface S of volume V,

$$\frac{\mathrm{d}}{\mathrm{d}t}\text{prob(in } V, t) = -\int_S \mathrm{d}\vec{S} \cdot \vec{j}(\vec{r}, t) \tag{3.1.4}$$

where $\vec{j}(\vec{r}, t)$ is the *probability current density*, and $\mathrm{d}\vec{S}$ is the outwardly-oriented vectorial surface element:

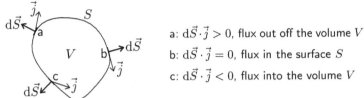

a: $d\vec{S}\cdot\vec{j} > 0$, flux out off the volume V

b: $d\vec{S}\cdot\vec{j} = 0$, flux in the surface S

c: $d\vec{S}\cdot\vec{j} < 0$, flux into the volume V

The minus sign in (3.1.4) accounts for the standard convention of regarding leaving flux as positive. The comparison of

$$\frac{\mathrm{d}}{\mathrm{d}t}\mathrm{prob}(\text{in }V,t) = \int_V (\mathrm{d}\vec{r})\,\frac{\partial}{\partial t}\rho(\vec{r},t) \tag{3.1.5}$$

with this consequence of Karl F. Gauss's theorem:

$$\int_S \mathrm{d}\vec{S}\cdot\vec{j}(\vec{r},t) = \int_V (\mathrm{d}\vec{r})\,\vec{\nabla}\cdot\vec{j}(\vec{r},t)\,, \tag{3.1.6}$$

then implies the *continuity equation*

$$\frac{\partial}{\partial t}\rho(\vec{r},t) + \vec{\nabla}\cdot\vec{j}(\vec{r},t) = 0\,, \tag{3.1.7}$$

the familiar local conservation law, here for probability.

We wish to establish an explicit expression for $\vec{j}(\vec{r},t)$ when the dynamics are of the typical simple kind, that is when the Hamilton operator has the standard form

$$H = \frac{1}{2M}\vec{P}^2 + V(\vec{R})\,. \tag{3.1.8}$$

What is important below is that the velocity is just proportional to the momentum

$$\frac{\mathrm{d}}{\mathrm{d}t}\vec{R}(t) = \frac{1}{M}\vec{P}(t)\,. \tag{3.1.9}$$

We proceed from

$$-\vec{\nabla}\cdot\vec{j}(\vec{r},t) = \frac{\partial}{\partial t}\rho(\vec{r},t) = \left\langle \frac{\mathrm{d}}{\mathrm{d}t}\delta\big(\vec{R}(t) - \vec{r}\big)\right\rangle \tag{3.1.10}$$

and use

$$\delta\big(\vec{R}(t) - \vec{r}\big) = \int \frac{(\mathrm{d}\vec{k})}{(2\pi)^3}\,\mathrm{e}^{\mathrm{i}\vec{k}\cdot(\vec{R}(t) - \vec{r})} \tag{3.1.11}$$

to carry out the time differentiation with the aid of the differentiation rule (1.4.41) for exponential functions for operators. This gives

$$
\frac{\mathrm{d}}{\mathrm{d}t}\delta\left(\vec{R}(t) - \vec{r}\right) = \int \frac{(\mathrm{d}\vec{k})}{(2\pi)^3}\int_0^1 \mathrm{d}x\; \mathrm{e}^{\mathrm{i}x\vec{k}\,\cdot\,(\vec{R}(t) - \vec{r})}\mathrm{i}\vec{k}\cdot\frac{\mathrm{d}\vec{R}(t)}{\mathrm{d}t}\;\mathrm{e}^{\mathrm{i}(1-x)\vec{k}\,\cdot\,(\vec{R}(t) - \vec{r})}
$$

$$
= \int \frac{(\mathrm{d}\vec{k})}{(2\pi)^3}\int_0^1 \mathrm{d}x\; \mathrm{e}^{\mathrm{i}x\vec{k}\,\cdot\,\vec{R}(t)}\mathrm{i}\vec{k}\cdot\frac{\vec{P}(t)}{M}\;\mathrm{e}^{-\mathrm{i}x\vec{k}\,\cdot\,\vec{R}(t)}\,\mathrm{e}^{\mathrm{i}\vec{k}\,\cdot\,(\vec{R}(t) - \vec{r})}.
$$

$$(3.1.12)$$

With

$$
\mathrm{e}^{\mathrm{i}x\vec{k}\,\cdot\,\vec{R}}\vec{P}\,\mathrm{e}^{-\mathrm{i}x\vec{k}\,\cdot\,\vec{R}} = \vec{P} - x\hbar\vec{k}\,, \tag{3.1.13}
$$

which is the three-dimensional version of the second statement in (1.2.71), we then get

$$
\frac{\mathrm{d}}{\mathrm{d}t}\delta\left(\vec{R}(t) - \vec{r}\right) = \frac{1}{M}\int \frac{(\mathrm{d}\vec{k})}{(2\pi)^3}\,\mathrm{i}\vec{k}\cdot\left(\vec{P}(t) - \frac{1}{2}\hbar\vec{k}\right)\mathrm{e}^{\mathrm{i}\vec{k}\,\cdot\,(\vec{R}(t) - \vec{r})} \tag{3.1.14}
$$

after evaluating the now elementary x integral. Equivalently, we can write

$$
\frac{\mathrm{d}}{\mathrm{d}t}\delta\left(\vec{R}(t) - \vec{r}\right) = \frac{1}{M}\int \frac{(\mathrm{d}\vec{k})}{(2\pi)^3}\,\mathrm{e}^{\mathrm{i}\vec{k}\,\cdot\,(\vec{R}(t) - \vec{r})}\mathrm{i}\vec{k}\cdot\left(\vec{P}(t) + \frac{1}{2}\hbar\vec{k}\right) \tag{3.1.15}
$$

or, taking half the sum of both for symmetrization,

$$
\frac{\mathrm{d}}{\mathrm{d}t}\delta\left(\vec{R}(t) - \vec{r}\right)
$$

$$
= \frac{1}{2M}\int \frac{(\mathrm{d}\vec{k})}{(2\pi)^3}\left(\mathrm{i}\vec{k}\cdot\vec{P}(t)\,\mathrm{e}^{\mathrm{i}\vec{k}\,\cdot\,(\vec{R}(t) - \vec{r})} + \mathrm{e}^{\mathrm{i}\vec{k}\,\cdot\,(\vec{R}(t) - \vec{r})}\mathrm{i}\vec{k}\cdot\vec{P}(t)\right)
$$

$$
= -\vec{\nabla}\cdot\frac{1}{2M}\int \frac{(\mathrm{d}\vec{k})}{(2\pi)^3}\left(\vec{P}(t)\,\mathrm{e}^{\mathrm{i}\vec{k}\,\cdot\,(\vec{R}(t) - \vec{r})} + \mathrm{e}^{\mathrm{i}\vec{k}\,\cdot\,(\vec{R}(t) - \vec{r})}\vec{P}(t)\right)
$$

$$
= -\vec{\nabla}\cdot\frac{1}{2M}\left[\vec{P}(t)\delta\left(\vec{R}(t) - \vec{r}\right) + \delta\left(\vec{R}(t) - \vec{r}\right)\vec{P}(t)\right] \tag{3.1.16}
$$

where the gradient differentiates with respect to the numerical vector $\vec{r}$.

Accordingly, we have

$$
\vec{j}(r,t) = \frac{1}{2M}\left\langle\left[\vec{P}(t)\delta\left(\vec{R}(t) - \vec{r}\right) + \delta\left(\vec{R}(t) - \vec{r}\right)\vec{P}(t)\right]\right\rangle \tag{3.1.17}
$$

for the probability current density when the dynamics is governed by a Hamilton operator of the typical form (3.1.8). Since

$$\vec{P}(t)\delta\left(\vec{R}(t) - \vec{r}\right) + \delta\left(\vec{R}(t) - \vec{r}\right)\vec{P}(t)$$
$$= \vec{P}(t)|\vec{r},t\rangle\langle\vec{r},t| + |\vec{r},t\rangle\langle\vec{r},t|\vec{P}(t)$$
$$= \left(i\hbar\vec{\nabla}|\vec{r},t\rangle\right)\langle\vec{r},t| + |\vec{r},t\rangle\frac{\hbar}{i}\vec{\nabla}\langle\vec{r},t| \qquad (3.1.18)$$

this becomes

$$\vec{j}(r,t) = \frac{1}{2M}\left(\left(i\hbar\vec{\nabla}\psi(\vec{r},t)^*\right)\psi(\vec{r},t) + \psi(\vec{r},t)^*\frac{\hbar}{i}\vec{\nabla}\psi(\vec{r},t)\right) \qquad (3.1.19)$$

if the system state is given by ket $|\ \rangle$ with position wave function $\psi(\vec{r},t) = \langle\vec{r},t|\ \rangle$, which appears more compactly when written in the form

$$\vec{j}(r,t) = \frac{1}{M}\,\text{Re}\left(\psi(\vec{r},t)^*\frac{\hbar}{i}\vec{\nabla}\psi(\vec{r},t)\right)$$
$$= \frac{\hbar}{M}\,\text{Im}\left(\psi(\vec{r},t)^*\vec{\nabla}\psi(\vec{r},t)\right). \qquad (3.1.20)$$

3-1   Use the Schrödinger equation that is obeyed by $\psi(\vec{r},t)$ to verify the continuity equation (3.1.7) for this $\vec{j}(r,t)$ and $\rho = |\psi(\vec{r},t)|^2$.

3-2   For a different derivation of (3.1.17), return to (3.1.10) and use the Heisenberg equation of motion (1.3.2) for

$$F\left(\underbrace{\vec{R}(t), \vec{P}(t), t}_{A(t)}\right) = \delta\left(\vec{R}(t) - \vec{r}\right) = |\vec{r},t\rangle\langle\vec{r},t|$$

with the Hamilton operator given in (3.1.8).

3-3   Show that

$$\frac{d}{dt}\int(d\vec{r})\,\vec{r}\rho(\vec{r},t) = \int(d\vec{r})\,\vec{j}(\vec{r},t).$$

3-4   For $H = \dfrac{1}{2M}\vec{P}^2$, force-free motion, and

$$\psi(\vec{r}, t = 0) = \pi^{-3/4} a^{-3/2}\, \mathrm{e}^{-\frac{1}{2}(r/a)^2}$$

with $a > 0$, determine the total flux through the sphere $r = s > 0$, as a function of time. Hint: If you remember what is said about the "spreading of the wave function" in *Basic Matters* and *Simple Systems*, you will not have much work.

3.2   One-dimensional prelude: Forces scatter

Let us now consider a one-dimensional situation with a localized potential energy,

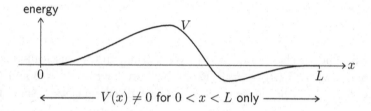

so that

$$H = \frac{1}{2M}P^2 + V(X)$$

$$\text{with}\quad V(x) = 0 \quad \text{for}\quad x < 0 \quad \text{and}\quad x > L\,. \tag{3.2.1}$$

We want to study a scattering situation, with incoming amplitudes from the left ($x < 0$) or the right ($x > L$), or from both sides. Sufficiently far to the left or right we have $H = P^2/(2M)$ essentially, so that the relevant eigenvalues of H are positive — or, put differently, we can ignore the bound states of H if there are any.

We parameterize the positive energies by $E = (\hbar k)^2/(2M)$ and thus have

$$\psi(x, t) = \int \mathrm{d}k\, A(k)\, \mathrm{e}^{-\mathrm{i}\frac{\hbar k^2}{2M}t}\phi(k, x) \tag{3.2.2}$$

for the general solution of Schrödinger's equation whereby $\phi(k, x)$ solves the time-independent Schrödinger equation, that is the eigenfunction equation

$$\left(-\frac{\hbar^2}{2M}\frac{\partial^2}{\partial x^2} + V(x)\right)\phi(k,x) = \frac{(\hbar k)^2}{2M}\phi(k,x)\,. \tag{3.2.3}$$

Equivalently, we write

$$\left[\frac{\partial^2}{\partial x^2} + k^2 - \frac{2M}{\hbar^2}V(x)\right]\phi(k,x) = 0 \tag{3.2.4}$$

or

$$\frac{\partial^2}{\partial x^2}\phi(k,x) = -k(x)^2\phi(k,x) \tag{3.2.5}$$

where

$$k(x) = \sqrt{k^2 - \frac{2M}{\hbar^2}V(x)}$$

is the *local* de Broglie wave number, $2\pi/k(x)$ being the local de Broglie wavelength, named after Prince Louis-Victor de Broglie.

In the classically allowed regions, where $V(x) < (\hbar k)^2/(2M) = E$, $k(x)^2$ is positive and we choose $k(x) > 0$ by convention. In the classically forbidden region, where $V(x) > (\hbar k)^2/(2M) = E$, we have $k(x)^2 < 0$ and $k(x)$ is imaginary. For the purpose of the present discussion, we assume that E is so large that $V(x) < E$ everywhere. The situation in which a classically allowed region is located between two classically forbidden regions is considered in Section 6.8 of *Simple Systems*.

Where $V(x) = 0$, we simply have $k(x) = k = \sqrt{2ME}/\hbar$ and

$$\phi(k,x) = \mathrm{e}^{\mathrm{i}kx} \quad \text{or} \quad \mathrm{e}^{-\mathrm{i}kx} \quad \text{for} \quad x < 0 \quad \text{or} \quad x > L\,. \tag{3.2.6}$$

The probability currents associated with these plane waves are

$$j \propto \mathrm{Im}\left(\mathrm{e}^{\mp\mathrm{i}kx}\frac{\partial}{\partial x}\mathrm{e}^{\pm\mathrm{i}kx}\right) = \pm k\,, \tag{3.2.7}$$

so that $\mathrm{e}^{+\mathrm{i}kx}$ would correspond to motion to the right (positive j) and $\mathrm{e}^{-\mathrm{i}kx}$ to motion to the left (negative j).

For $k(x) > 0$, the systematic decomposition of the general $\phi(k,x)$ into left and right moving parts is achieved by putting

$$\phi(k,x) = \frac{1}{\sqrt{k(x)}}\left[\phi_+(k,x) + \phi_-(k,x)\right]\,,$$

$$\frac{1}{\mathrm{i}}\frac{\partial}{\partial x}\phi(k,x) = \sqrt{k(x)}\left[\phi_+(k,x) - \phi_-(k,x)\right]\,, \tag{3.2.8}$$

because then we have

$$
\begin{aligned}
j(x) &= \frac{\hbar}{M} \operatorname{Re}\left(\phi(k,x)^* \frac{1}{i} \frac{\partial}{\partial x} \phi(k,x) \right) \\
&= \frac{\hbar}{M} \operatorname{Re}\left((\phi_+ + \phi_-)^*(\phi_+ - \phi_-) \right) \\
&= \frac{\hbar}{M} \left(|\phi_+|^2 - |\phi_-|^2 \right).
\end{aligned}
\tag{3.2.9}
$$

As we see, ϕ_+ *always* gives a positive contribution to $j(x)$, whereas ϕ_- always gives a negative contribution, and this justifies our interpretation that ϕ_+ is the component moving from left to right and ϕ_- the component that moves from right to left.

3-5 How would you have to modify (3.2.8) and (3.2.9) for $k(x)^2 < 0$?

The equations obeyed by $\phi_\pm(k,x)$ follow from the Schrödinger equation for $\phi = (\phi_+ + \phi_-)/\sqrt{k(x)}$. We compare

$$
\begin{aligned}
&\left[i\frac{\partial}{\partial x} \pm k(x) \right] \left[\frac{1}{i}\frac{\partial}{\partial x} \pm k(x) \right] \phi(k,x) \\
&= \left[i\frac{\partial}{\partial x} \pm k(x) \right] \left(\pm 2\sqrt{k(x)} \right) \phi_\pm(k,x) \\
&= \pm 2\sqrt{k(x)} \left(i\frac{\partial}{\partial x} \pm k(x) + \frac{i}{2}\frac{1}{k(x)}\frac{\partial k(x)}{\partial x} \right) \phi_\pm(k,x) \quad (3.2.10)
\end{aligned}
$$

with

$$
\begin{aligned}
&\left[i\frac{\partial}{\partial x} \pm k(x) \right] \left[\frac{1}{i}\frac{\partial}{\partial x} \pm k(x) \right] \phi(k,x) \\
&= \left(\frac{\partial^2}{\partial x^2} + k(x)^2 \pm i\frac{\partial k(x)}{\partial x} \right) \phi(k,x) \\
&= \pm i\frac{\partial k(x)}{\partial x} \phi(k,x) \\
&= \pm i\frac{\partial k(x)}{\partial x} \frac{1}{\sqrt{k(x)}} (\phi_+(k,x) + \phi_-(k,x)) \tag{3.2.11}
\end{aligned}
$$

to establish

$$
\left[i\frac{\partial}{\partial x} \pm k(x) \right] \phi_\pm(k,x) = \frac{i}{2}\frac{1}{k(x)}\frac{\partial k(x)}{\partial x} \phi_\mp(k,x).
\tag{3.2.12}
$$

We note that $k(x)^2 = k^2 - \dfrac{2M}{\hbar^2} V(x)$ implies

$$2k(x)\frac{\partial}{\partial x}k(x) = -\frac{2M}{\hbar^2}\frac{\partial V}{\partial x} = \frac{2M}{\hbar^2}F(x) \qquad (3.2.13)$$

or

$$\frac{1}{k(x)}\frac{\partial}{\partial x}k(x) = \frac{(M/\hbar^2)F(x)}{k(x)^2} = \frac{1}{2}\frac{F(x)}{\frac{\hbar^2}{2M}k(x)^2}$$

$$= \frac{1}{2}\frac{F(x)}{E - V(x)}, \qquad (3.2.14)$$

so that

$$\left[i\frac{\partial}{\partial x} \pm k(x)\right]\phi_{\pm}(k, x) = \frac{i}{4}\frac{F(x)}{E - V(x)}\phi_{\mp}(k, x). \qquad (3.2.15)$$

This says that the *force* $F(x) = -\dfrac{\partial}{\partial x}V(x)$ is responsible for feeding the left-moving amplitude ϕ_- into the right-moving amplitude ϕ_+ and vice versa — the force gives rise to scattering events.

If the force is small, that is

$$\left|\frac{F(x)}{E - V(x)}\right| \ll k(x) = \sqrt{\frac{2M}{\hbar}\left[E - V(x)\right]} \qquad (3.2.16)$$

or

$$\left|\frac{\partial}{\partial x}\frac{1}{k(x)}\right| \ll 1, \qquad (3.2.17)$$

then we can neglect the right-hand side in (3.2.15) and have the simple approximative solutions

$$\phi_+(k, x) = \phi_+(k, 0)\,e^{\displaystyle i\int_0^x dx'\,k(x')}\,,$$

$$\phi_-(k, x) = \phi_-(k, L)\,e^{\displaystyle i\int_x^L dx'\,k(x')}\,, \qquad (3.2.18)$$

for which

$$\phi_+(k, L) = e^{i\alpha}\phi_+(k, 0)\,,$$

$$\phi_-(k, 0) = e^{i\alpha}\phi_-(k, L)\,, \qquad (3.2.19)$$

with the accumulated phase α given by

$$\alpha = \int_0^L \mathrm{d}x\, k(x)\,. \tag{3.2.20}$$

More generally, the "in" amplitudes $\phi_+(k,0), \phi_-(k,L)$ lead to "out" amplitudes $\phi_+(k,L), \phi_-(k,0)$, that are given by a linear transformation,

$$\begin{pmatrix} \phi_+(k,L) \\ \phi_-(k,0) \end{pmatrix} = \mathrm{e}^{\mathrm{i}\alpha} \begin{pmatrix} S_{++} & S_{+-} \\ S_{-+} & S_{--} \end{pmatrix} \begin{pmatrix} \phi_+(k,0) \\ \phi_-(k,L) \end{pmatrix}, \tag{3.2.21}$$

pictorially:

in $\phi_+(k,0) \longrightarrow$ $\longrightarrow \phi_+(k,L)$ out

out $\phi_-(k,0) \longleftarrow$ $\longleftarrow \phi_-(k,L)$ in

The 2×2 matrix S that turns the "in" amplitudes into the "out" amplitudes is the *scattering matrix* for this simple example.

3-6 Show that this scattering matrix is unitary as a consequence of the continuity equation (3.1.7).

3-7 Write

$$\begin{pmatrix} S_{++} & S_{+-} \\ S_{-+} & S_{--} \end{pmatrix} = \begin{pmatrix} 1 & 0 \\ 0 & 1 \end{pmatrix} - \mathrm{i} \begin{pmatrix} K_{++} & K_{+-} \\ K_{-+} & K_{--} \end{pmatrix}$$

and determine K_{++}, K_{+-}, K_{-+}, and K_{--} to first order in $\dfrac{F(x)}{E - V(x)}$.

3-8 Use an idea from page 50 to turn this into a *unitary* approximation for the scattering matrix.

3.3 Scattering by a localized potential

3.3.1 *Golden-rule approximation*

In the typical three-dimensional scattering situation, we have a localized potential $V(\vec{r})$ and particles (atoms, electrons, ...) approaching with rather well defined momentum:

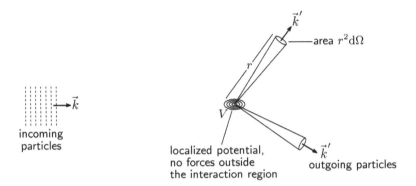

The total Hamilton operator is

$$H = \underbrace{\frac{1}{2M}\vec{P}^2}_{= H_0} + \underbrace{V(\vec{R})}_{= H_1} \qquad (3.3.1)$$

with a break-up into "unperturbed motion" governed by H_0, the kinetic energy $\vec{P}^2/2M$, and the perturbation H_1, the localized potential energy $V(\vec{R})$.

Although not really necessary, we shall usually regard the scattering potential as a function of the position operator only, $V = V(\vec{R})$, but there are more general cases in which a momentum dependence is present as well. The separable potential of Exercise 3-9 on page 99 below is an example. Most results do not rely on the assumption that V does not depend on $\vec{P}$, and the others are generalized rather easily.

The forces resulting from the potential $V(\vec{R})$ deflect the particle, which is to say that they induce transitions between the eigenstates of H_0, namely the states with definite momentum:

$$\text{initial state} \quad |2\rangle = |\vec{k}\rangle, \quad \vec{P}|\vec{k}\rangle = |\vec{k}\rangle\hbar\vec{k}\,,$$

$$\text{final state} \quad \langle 1| = \langle \vec{k}'|, \quad \langle \vec{k}'|\vec{P} = \hbar\vec{k}'\langle \vec{k}'|\,,$$

$$H_0|2\rangle = |2\rangle\frac{(\hbar k)^2}{2M} \quad \text{with} \quad k = |\vec{k}|\,,$$

$$\langle 1|H_0 = \frac{(\hbar k')^2}{2M}\langle 1| \quad \text{with} \quad k' = |\vec{k}'|\,. \qquad (3.3.2)$$

We apply the golden rule of Section 2.4 and get

$$\text{transition rate} = \frac{2\pi}{\hbar} \left| \langle \vec{k}' | V | \vec{k} \rangle \right|^2 \delta \left(\frac{(\hbar k')^2}{2M} - \frac{(\hbar k)^2}{2M} \right)$$

(and summation over all final states) (3.3.3)

where the required summation is over all $\vec{k}'$ that point into the same direction, specified by the solid-angle element $d\Omega$,

$$\text{summation} \stackrel{\wedge}{=} d\Omega \int_0^\infty dk' \, k'^2 . \tag{3.3.4}$$

Thus

$$\text{transition rate} = d\Omega \frac{2\pi}{\hbar} \int_0^\infty dk' \, k'^2 \delta \left(\frac{(\hbar k')^2}{2M} - \frac{(\hbar k)^2}{2M} \right) \left| \langle \vec{k}' | V | \vec{k} \rangle \right|^2 .$$

(3.3.5)

It is common and convenient and physically very meaningful, to normalize the rate to the incoming flux, because the number of scattering events per second depends on the scattering power of the potential but is also proportional to the number of particles approaching per second. To extract the scattering power of the potential we therefore write

$$\text{transition rate} = (\text{incoming flux}) \times \frac{d\sigma}{d\Omega} d\Omega \tag{3.3.6}$$

and thus introduce the *differential scattering cross section* $\frac{d\sigma}{d\Omega}$. It is the effective transverse area of the target for scattering into the specified direction.

With the wave function

$$\psi_{\text{in}}(\vec{r}) = \langle \vec{r} | \vec{k} \rangle = \frac{1}{(2\pi)^{3/2}} e^{i\vec{k} \cdot \vec{r}} , \tag{3.3.7}$$

the incoming flux is the magnitude of the probability current

$$\vec{j}_{\text{in}} = \frac{\hbar}{M} \text{Im} \left(\psi_{\text{in}}^* \vec{\nabla} \psi_{\text{in}} \right) = \frac{\hbar}{M} \frac{1}{(2\pi)^3} \vec{k} , \tag{3.3.8}$$

that is

$$\text{incoming flux} = \frac{\hbar}{M} \frac{k}{(2\pi)^3} . \tag{3.3.9}$$

Accordingly,

$$\frac{\mathrm{d}\sigma}{\mathrm{d}\Omega} = \frac{M}{\hbar k}\frac{(2\pi)^4}{\hbar}\int_0^\infty \mathrm{d}k'\, k'^2 \left|\langle \vec{k}'|V|\vec{k}\rangle\right|^2 \delta\!\left(\frac{\hbar^2}{2M}(k'^2 - k^2)\right) \qquad (3.3.10)$$

is the golden-rule approximation for $\dfrac{\mathrm{d}\sigma}{\mathrm{d}\Omega}$. We note that the δ-function factor enforces *elastic* scattering here, that is: the scattered particle has the same kinetic energy as the incoming particle, no energy is transferred to the particle.

Inelastic scattering can also happen, of course, for example when an electron is scattered by an atom and excites the atom during the scattering act. The outgoing electron would then have less kinetic energy than the incoming one, less by the amount needed for the excitation of the atom. You can easily envision even more complicated situations for yourself.

After writing

$$\delta\!\left(\frac{\hbar^2}{2M}(k'^2 - k^2)\right) = \frac{2M}{\hbar^2}\delta\!\left(k'^2 - k^2\right)$$

$$= \frac{M}{\hbar^2}\frac{1}{k}\delta(k' - k) \qquad (3.3.11)$$

we get ready for the evaluation of the k' integral in (3.3.10) which yields

$$\frac{\mathrm{d}\sigma}{\mathrm{d}\Omega} = \frac{(2\pi)^4 M^2}{\hbar^4}\left|\langle \vec{k}'|V|\vec{k}\rangle\right|^2\bigg|_{k'=k}, \qquad (3.3.12)$$

the golden-rule approximation for the differential scattering cross section $\dfrac{\mathrm{d}\sigma}{\mathrm{d}\Omega}$. If $V = V(\vec{R})$, the transition matrix element $\langle \vec{k}'|V|\vec{k}\rangle$ can be expressed as

$$\langle \vec{k}'|V|\vec{k}\rangle = \int (\mathrm{d}\vec{r})\,\langle \vec{k}'|\vec{r}\rangle V(\vec{r})\langle \vec{r}|\vec{k}\rangle$$

$$= \frac{1}{(2\pi)^3}\int (\mathrm{d}\vec{r})\,V(\vec{r})\,\mathrm{e}^{\mathrm{i}(\vec{k}-\vec{k}')\cdot\vec{r}} \qquad (3.3.13)$$

so that

$$\frac{\mathrm{d}\sigma}{\mathrm{d}\Omega} = \left|\frac{M}{2\pi\hbar^2}\int (\mathrm{d}\vec{r})\,V(\vec{r})\,\mathrm{e}^{\mathrm{i}(\vec{k}-\vec{k}')\cdot\vec{r}}\right|^2\bigg|_{k'=k} \qquad (3.3.14)$$

is the golden-rule approximation for $\dfrac{\mathrm{d}\sigma}{\mathrm{d}\Omega}$ in the case of elastic potential scattering by $V = V(\vec{R})$.

3-9   In nuclear physics one sometimes approximates the complicated forces by a *separable potential*, that is

$$V = |s\rangle E_0 \langle s| \,,$$

which is essentially the projector $|s\rangle\langle s|$ for some appropriate ket $|s\rangle$ and its bra $\langle s|$. What is $\dfrac{\mathrm{d}\sigma}{\mathrm{d}\Omega}$ in this case?

A simplification occurs in the important situation of a spherically symmetric potential,

$$V(\vec{r}) = V(r) \,, \tag{3.3.15}$$

because then the dependence on the direction of the integration variable $\vec{r}$ is solely in the plane-wave factor

$$e^{i(\vec{k}-\vec{k}')\cdot\vec{r}} = e^{i|\vec{k}-\vec{k}'|r\cos\vartheta}$$
$$= e^{iqr\cos\vartheta} \quad \text{with} \quad q = |\vec{k}-\vec{k}'| \,. \tag{3.3.16}$$

We take $\vec{k}-\vec{k}'$ as the direction of the polar axis, the z axis of spherical coordinates, so that

$$\int (\mathrm{d}\vec{r}) \to 2\pi \int_0^\infty \mathrm{d}r\, r^2 \int_0^\pi \mathrm{d}\vartheta\, \sin\vartheta \,, \tag{3.3.17}$$

and arrive at

$$\int (\mathrm{d}\vec{r})\, V(\vec{r})\, e^{i(\vec{k}-\vec{k}')\cdot\vec{r}} = 2\pi \int_0^\infty \mathrm{d}r\, r^2 V(r) \int_0^\pi \mathrm{d}\vartheta\, \sin\vartheta\, e^{iqr\cos\vartheta}$$

$$= 2\pi \int_0^\infty \mathrm{d}r\, r^2 V(r) \frac{e^{iqr\cos\vartheta}}{-iqr}\bigg|_{\vartheta=0}^{\vartheta=\pi}$$

$$= 4\pi \frac{1}{q} \int_0^\infty \mathrm{d}r\, r V(r) \sin(qr) \,, \tag{3.3.18}$$

with the consequence

$$\frac{\mathrm{d}\sigma}{\mathrm{d}\Omega} = \left| \frac{2M}{\hbar^2 q} \int_0^\infty \mathrm{d}r\, r V(r) \sin(qr) \right|^2 \tag{3.3.19}$$

where

$$q = |\vec{k}-\vec{k}'| = 2k\sin\frac{\theta}{2} \tag{3.3.20}$$

identifies the deflection angle θ in accordance with

$$\vec{k} \cdot \vec{k}' = k^2 \cos\theta. \qquad (3.3.21)$$

3.3.2 Example: Yukawa potential

As an illustrating example, we consider the so-called Yukawa potential, named after Hideki Yukawa,

$$V(r) = \frac{V_0}{\kappa r} e^{-\kappa r}, \qquad (3.3.22)$$

which is a shielded Coulomb potential with the range $1/\kappa$ determined by the parameter κ. We evaluate the required integral

$$
\begin{aligned}
\int_0^\infty \mathrm{d}r\, r V(r) \sin(qr) &= \frac{V_0}{\kappa} \int_0^\infty \mathrm{d}r\, e^{-\kappa r} \sin(qr) \\
&= \frac{V_0}{\kappa} \operatorname{Im}\left(\int_0^\infty \mathrm{d}r\, e^{-\kappa r + iqr} \right) \\
&= \frac{V_0}{\kappa} \operatorname{Im}\left(\frac{1}{\kappa - iq} \right) = \frac{V_0}{\kappa} \frac{q}{\kappa^2 + q^2}, \quad (3.3.23)
\end{aligned}
$$

and so get

$$\frac{\mathrm{d}\sigma}{\mathrm{d}\Omega} = \left| \frac{2M}{\hbar^2 q} \frac{V_0}{\kappa} \frac{q}{\kappa^2 + q^2} \right|^2 = \left| \frac{2MV_0/\kappa}{\hbar^2(\kappa^2 + q^2)} \right|^2. \qquad (3.3.24)$$

The combination of parameters

$$
\begin{aligned}
\frac{\hbar^2}{2M} q^2 &= \frac{\hbar^2}{2M}\left(2k \sin\frac{\theta}{2} \right)^2 = \frac{(\hbar k)^2}{2M} 4\left(\sin\frac{\theta}{2} \right)^2 \\
&= 4E\left(\sin\frac{\theta}{2} \right)^2
\end{aligned}
\qquad (3.3.25)
$$

involves the energy of the scattered particle, and this invites us to associate an energy scale

$$E_0 = \frac{(\hbar\kappa)^2}{2M} \qquad (3.3.26)$$

with the range parameter κ. Then

$$\frac{d\sigma}{d\Omega} = \left(\frac{V_0/\kappa}{E_0 + 4E\left(\sin\frac{\theta}{2}\right)^2} \right)^2 \tag{3.3.27}$$

is the compact result for $\frac{d\sigma}{d\Omega}$. For large energy E, that is large kinetic energy of the scattered particles, the differential cross section $\frac{d\sigma}{d\Omega} \propto (1/E)^2$ is small, as could be expected on physical grounds: It is more difficult to deflect a fast moving object than a slow moving one.

3-10 Determine the total scattering cross section

$$\sigma = \int d\Omega \, \frac{d\sigma}{d\Omega} = 2\pi \int_0^\pi d\theta \, \sin\theta \frac{d\sigma}{d\Omega} \,.$$

How does it scale with E for $E \gg E_0$?

3-11 By

$$\overline{\cos\theta} = \frac{\displaystyle\int d\Omega \, \frac{d\sigma}{d\Omega} \cos\theta}{\displaystyle\int d\Omega \, \frac{d\sigma}{d\Omega}}$$

one can define an average value of $\cos\theta$. How large is it for the Yukawa potential? When $\overline{\cos\theta} \lesssim 1$, we can write $\overline{\cos\theta} \cong 1 - \frac{1}{2}\overline{\theta^2}$. Find $\overline{\theta^2}$ for $E \gg E_0$.

3-12 Consider the scattering by the double Yukawa potential

$$V(\vec{r}) = V(\vec{r} - \vec{a}) + V(\vec{r} + \vec{a}) \quad \text{with} \quad V(\vec{r}) = \frac{V_0}{\kappa r}\,e^{-\kappa r} \,,$$

where $\vec{a}$ is parallel to $\vec{k}$, that is $\vec{k}\cdot\vec{a} = ka > 0$. Find the scattering amplitude $f(\theta)$ and the differential cross section $\frac{d\sigma}{d\Omega}(\theta)$ in Born approximation. If it is observed that no scattering occurs in the three directions for which $\cos\theta = 0$ or $\cos\theta = \pm\frac{2}{3}$, how big is the spacing a between the scattering centers?

3.3.3 *Rutherford cross section as a limit*

The Coulomb potential

$$V = -\frac{Ze^2}{r} \,, \tag{3.3.28}$$

experienced by an electron (charge $-e$) in the electrostatic field of a nucleus (charge Ze) can be regarded as the limit of the Yukawa potential in the sense of

$$\kappa \to 0 \quad \text{with} \quad V_0/\kappa \to -Ze^2 \,. \tag{3.3.29}$$

Then $E_0 \to 0$ in (3.3.26) and (3.3.27) and

$$\frac{\mathrm{d}\sigma}{\mathrm{d}\Omega} \to \left(\frac{Ze^2}{4E\left(\sin\frac{\theta}{2}\right)^2} \right)^2 = \frac{1}{16} \left(\frac{Ze^2}{E} \right)^2 \frac{1}{\left(\sin\frac{\theta}{2}\right)^4} \,. \tag{3.3.30}$$

This is the famous *Rutherford cross section* for Coulomb scattering, named in honor of Ernest Rutherford, Lord Rutherford of Nelson.

By a remarkable coincidence, we get the exact Rutherford cross section by the approximate golden-rule treatment, despite the fact that the formalism should not really apply to potentials with long-range forces such as the Coulomb potential. Equally remarkable is the fact that this quantum mechanical cross section agrees perfectly with its classical analog. Actually, Rutherford derived it by purely classical arguments, that is: he applied Isaac Newton's classical mechanics to the problem rather than the quantum mechanics that was only developed more than a decade later.

3.4   Lippmann–Schwinger equation

In the general scattering situation

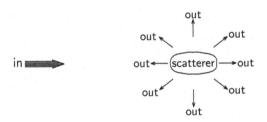

we characterize the "in" state as an eigenstate of the Hamilton operator H_0 of the unperturbed evolution,

$$H_0|\text{in}\rangle = |\text{in}\rangle E \,, \tag{3.4.1}$$

whereas the "out" state is an eigenstate of the full Hamilton operator to the *same* energy,

$$H|\text{out}\rangle = |\text{out}\rangle E, \quad H = H_0 + V, \qquad (3.4.2)$$

so that $|\text{out}\rangle$ contains both $|\text{in}\rangle$ and the scatterer's contribution. The scattering potential $V = H - H_0$ is meant to be localized, or more precisely: it does not have noticeable effects at large distances. This property of V is quite essential because without it we could not possible have such "in" states.

We are interested in the "out" state that is associated with the given "in" state and, therefore, we require

$$|\text{out}\rangle \to |\text{in}\rangle \quad \text{for} \quad V \to 0. \qquad (3.4.3)$$

We have

$$V|\text{out}\rangle = (E - H_0)|\text{out}\rangle = (E - H_0)(|\text{out}\rangle - |\text{in}\rangle) \qquad (3.4.4)$$

because $(E - H_0)|\text{in}\rangle = 0$. Now solving for $|\text{out}\rangle - |\text{in}\rangle$, the scatterer's contribution to $|\text{out}\rangle$, we get

$$|\text{out}\rangle - |\text{in}\rangle = (E - H_0)^{-1}V|\text{out}\rangle \qquad (3.4.5)$$

or

$$|\text{out}\rangle = |\text{in}\rangle + (E - H_0)^{-1}V|\text{out}\rangle. \qquad (3.4.6)$$

This has $|\text{out}\rangle = |\text{in}\rangle$ for $V = 0$ built into it all right, but it is suffering from a fundamental deficiency: the inverse operator $(E - H_0)^{-1}$ does not exist!

We encounter essentially the same problem in the perturbation theory for ground states, where we resolve it by restricting the inverse to the subspace orthogonal to the unperturbed state. This is the function of the operator Q_n in Section 6.5 of *Simple Systems*. This remedy is, however, not available in the present context because there is a continuum of scattering states, so that we are in trouble even if we manage to remove $|\text{in}\rangle\langle\text{in}|$ from $E - H_0$. Other states have energies arbitrarily close to E and render $(E - H_0)^{-1}$ singular, and then there are, of course, all the other "in" states with exactly the same energy as the $|\text{in}\rangle$ under consideration, differing from it by other quantum numbers, such as those specifying the direction of the approach.

We must fix the problem with another tool, which is the addition of a small imaginary part to $E - H_0$, after which it does have an inverse,

$$|\text{out}_\pm\rangle = |\text{in}\rangle + (E - H_0 \pm i\epsilon)^{-1} V |\text{out}_\pm\rangle \qquad (3.4.7)$$

with $0 < \epsilon \to 0$ eventually, at a stage where there is no longer any ambiguity. Presently we cannot decide whether a positive imaginary part or a negative one is the right choice, or whether it is necessary (it is not!) to consider more complicated ways of performing the limit $\epsilon \to 0$. Therefore, we postpone this decision about the sign in (3.4.7) until we will have learned enough about its right-hand side.

Let us turn the ket equation (3.4.7) into an equation for the wave functions

$$\psi_\pm(\vec{r}) = \langle \vec{r} | \text{out}_\pm \rangle \quad \text{and} \quad \phi(\vec{r}) = \langle \vec{r} | \text{in} \rangle, \qquad (3.4.8)$$

that is

$$\psi_\pm(\vec{r}) = \phi(\vec{r}) + \int (\mathrm{d}\vec{r}') \, \langle \vec{r} | (E - H_0 \pm i\epsilon)^{-1} | \vec{r}' \rangle \langle \vec{r}' | V | \text{out}_\pm \rangle. \qquad (3.4.9)$$

Upon assuming, solely for the sake of simplicity, a potential energy that only depends on the position $\vec{R}$ but not on the momentum $\vec{P}$,

$$\langle \vec{r}' | V(\vec{R}) = V(\vec{r}') \langle \vec{r}' |, \qquad (3.4.10)$$

we have

$$\psi_\pm(\vec{r}) = \phi(\vec{r}) + \int (\mathrm{d}\vec{r}') \, \langle \vec{r} | (E - H_0 \pm i\epsilon)^{-1} | \vec{r}' \rangle V(\vec{r}') \psi_\pm(\vec{r}'), \qquad (3.4.11)$$

where the dependence on the scattering potential is more explicit. The kernel in both the general (3.4.9) and the more specific (3.4.11),

$$\langle \vec{r} | (E - H_0 \pm i\epsilon)^{-1} | \vec{r}' \rangle \equiv G_\pm(\vec{r}, \vec{r}'), \qquad (3.4.12)$$

is a Green's function (George Green) by its nature and summarizes the net propagation effect of the unperturbed evolution.

In the most important, because most frequent, situation of

$$H_0 = \frac{1}{2M} \vec{P}^2 \qquad (3.4.13)$$

we can evaluate $G_\pm(\vec{r}, \vec{r}')$ quite explicitly. For this purpose we introduce momentum eigenstates to write

$$(E - H_0 \pm i\epsilon)^{-1} = \int (d\vec{p}) \, |\vec{p}\rangle \left(E - \frac{\vec{p}^2}{2M} \pm i\epsilon \right)^{-1} \langle \vec{p}|, \qquad (3.4.14)$$

and with

$$\langle \vec{r}|\vec{p}\rangle\langle \vec{p}|\vec{r}'\rangle = \frac{e^{i(\vec{r} - \vec{r}') \cdot \vec{p}/\hbar}}{(2\pi\hbar)^3} \qquad (3.4.15)$$

we then get

$$G_\pm(\vec{r}, \vec{r}') = \int \frac{(d\vec{p})}{(2\pi\hbar)^3} \frac{e^{i(\vec{r} - \vec{r}') \cdot \vec{p}/\hbar}}{E - \frac{\vec{p}^2}{2M} \pm i\epsilon} \qquad (3.4.16)$$

for the Green's function. The denominator depends only on the length $|\vec{p}|$ of the momentum variable $\vec{p}$, so that all directional dependence is contained in

$$(\vec{r} - \vec{r}') \cdot \vec{p} = |\vec{r} - \vec{r}'||\vec{p}| \cos\theta \qquad (3.4.17)$$

where θ is the angle between $\vec{r} - \vec{r}'$ and $\vec{p}$. As we did in (3.3.17), we choose the coordinate system such that $\vec{r} - \vec{r}'$ is along the z axis and then use spherical coordinates for the parameterization of the integration,

$$(d\vec{p}) = 2\pi\hbar^3 d\kappa \kappa^2 d\theta \sin\theta \quad \text{with} \quad \hbar\kappa = |\vec{p}|$$
$$\text{and} \quad (\vec{r} - \vec{r}') \cdot \vec{p} = \hbar\kappa s \cos\theta \quad \text{with} \quad s \equiv |\vec{r} - \vec{r}'|, \qquad (3.4.18)$$

where the lack of azimuthal dependence gives us the factor of 2π. Thus,

$$\begin{aligned}
G_\pm(\vec{r}, \vec{r}') &= \frac{1}{(2\pi)^2} \int_0^\infty d\kappa \, \kappa^2 \int_0^\pi d\theta \sin\theta \frac{e^{i\kappa s \cos\theta}}{E - \frac{(\hbar\kappa)^2}{2M} \pm i\epsilon} \\
&= \frac{1}{(2\pi)^2} \int_0^\infty d\kappa \, \kappa^2 \frac{\kappa^2}{E - \frac{(\hbar\kappa)^2}{2M} \pm i\epsilon} \frac{e^{i\kappa s \cos\theta}}{-i\kappa s} \Bigg|_{\theta = 0}^\pi \\
&= \frac{1}{(2\pi)^2} \frac{2}{s} \int_0^\infty d\kappa \frac{\kappa \sin(\kappa s)}{E - \frac{(\hbar\kappa)^2}{2M} \pm i\epsilon}.
\end{aligned} \qquad (3.4.19)$$

We now write the denominator as

$$\frac{\hbar^2}{2M}\left(\underbrace{\frac{2ME}{\hbar^2}}_{=\,k^2,\,k>0}-\kappa^2\pm i\epsilon\right)=\frac{\hbar^2}{2M}\left((k\pm i\epsilon)^2-\kappa^2\right)$$

$$=-\frac{\hbar^2}{2M}\left[\kappa-(k\pm i\epsilon)\right]\left[\kappa+(k\pm i\epsilon)\right],\qquad(3.4.20)$$

keeping in mind that ϵ is just an infinitesimal positive quantity. We note that the integrand is even in κ, which allows the extension of the integration range $0\cdots\infty$ to $-\infty\cdots\infty$, so that

$$\begin{aligned}
G_\pm\left(\vec{r},\vec{r}'\right)&=\frac{1}{(2\pi)^2}\left(-\frac{2M}{\hbar^2}\right)\frac{1}{s}\int_{-\infty}^{\infty}d\kappa\,\frac{\kappa\sin(\kappa s)}{\left[\kappa-(k\pm i\epsilon)\right]\left[\kappa+(k\pm i\epsilon)\right]}\\
&=\frac{1}{(2\pi)^2}\left(-\frac{2M}{\hbar^2}\right)\frac{1}{is}\int_{-\infty}^{\infty}d\kappa\,\frac{\kappa\,e^{i\kappa s}}{\left[\kappa-(k\pm i\epsilon)\right]\left[\kappa+(k\pm i\epsilon)\right]}\\
&=\frac{1}{(2\pi)^2}\left(-\frac{2M}{\hbar^2}\right)\frac{1}{2is}\int_{-\infty}^{\infty}d\kappa\left(\frac{e^{i\kappa s}}{\kappa-(k\pm i\epsilon)}+\frac{e^{i\kappa s}}{\kappa+(k\pm i\epsilon)}\right).
\end{aligned}$$

$$(3.4.21)$$

In the second step we replaced $\sin(\kappa s)\to\frac{1}{i}e^{i\kappa s}$, which is permissible because the additional $\frac{1}{i}\cos(\kappa s)$ term does not contribute to the value of the integral because it is multiplied by a function that is odd in κ.

The remaining integral is a standard example for contour integration with the aid of residues. In fact, a very similar integral appears in Section 5.4 of *Basic Matters*. The integration along the k axis is closed by a half-circle in the upper half-plane of the complex κ plane:

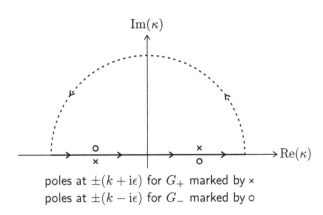

poles at $\pm(k+i\epsilon)$ for G_+ marked by ×
poles at $\pm(k-i\epsilon)$ for G_- marked by ○

We pick up the residue at $\kappa=k+i\epsilon$ for G_+ and that at $\kappa=-k+i\epsilon$ for

G_-, so that

$$G_\pm(\vec{r},\vec{r}') = \frac{1}{(2\pi)^2}\left(-\frac{2M}{\hbar^2}\right)\frac{1}{2is}2\pi i\, e^{i(\pm k\,+\,i\epsilon)s}$$

$$= -\frac{M}{2\pi\hbar^2}\frac{e^{\pm iks}}{s}\,e^{-\epsilon s}. \tag{3.4.22}$$

It is now safe to perform the $\epsilon \to 0$ limit, and

$$G_\pm(\vec{r},\vec{r}') = -\frac{M}{2\pi\hbar^2}\frac{e^{\pm ik|\vec{r}-\vec{r}'|}}{|\vec{r}-\vec{r}'|} \tag{3.4.23}$$

with $k = \sqrt{2ME/\hbar^2}$ is the outcome.

We have then a more explicit version of (3.4.11),

$$\psi_\pm(\vec{r}) = \phi(\vec{r}) - \frac{M}{2\pi\hbar^2}\int (d\vec{r}')\frac{e^{\pm ik|\vec{r}-\vec{r}'|}}{|\vec{r}-\vec{r}'|}V(\vec{r}')\psi_\pm(\vec{r}'), \tag{3.4.24}$$

and can finally address the question of the correct choice of sign in (3.4.7). For this purpose, we consider the asymptotic form, the large-r version, of $G_\pm(\vec{r}-\vec{r}')$, which involves

$$|\vec{r}-\vec{r}'| = \sqrt{r^2 + r'^2 - 2\vec{r}\cdot\vec{r}'}$$

$$= r\sqrt{1 - 2\frac{\vec{r}}{r}\cdot\vec{r}'/r + (r'/r)^2}$$

$$= r\left(1 - \frac{\vec{r}}{r}\cdot\vec{r}'/r + O\big((r'/r)^2\big)\right)$$

$$= r - \frac{\vec{r}}{r}\cdot\vec{r}' + O\big(r'^2/r\big). \tag{3.4.25}$$

The integration variable $\vec{r}'$ is limited to the scattering region, where $V(\vec{r}') \neq 0$, so that r'/r is negligibly small for large r:

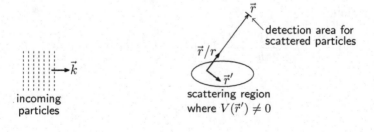

Thus, the asymptotic form of $G_\pm$ is of

$$G_\pm\left(\vec{r},\vec{r}'\right) \cong -\frac{M}{2\pi\hbar^2}\frac{\mathrm{e}^{\pm\mathrm{i}kr}}{r}\,\mathrm{e}^{\mp\mathrm{i}k\frac{\vec{r}}{r}\cdot\vec{r}'} \qquad (3.4.26)$$

and the scattered wave function is

$$\psi_{\pm,\mathrm{scat}}(\vec{r}) = -\frac{M}{2\pi\hbar^2}\frac{\mathrm{e}^{\pm\mathrm{i}kr}}{r}\int\left(\mathrm{d}\vec{r}'\right)\,\mathrm{e}^{\mp\mathrm{i}k\frac{\vec{r}}{r}\cdot\vec{r}'}V(\vec{r}')\psi_\pm(\vec{r}')\,. \qquad (3.4.27)$$

Its r dependence is dominated by the prefactor $\mathrm{e}^{\pm\mathrm{i}kr}/r$, which gives rise to a probability current of

$$\vec{\jmath}_{\mathrm{scat}} \propto \mathrm{Im}\left(\frac{1}{r}\,\mathrm{e}^{\mp\mathrm{i}kr}\vec{\nabla}\frac{1}{r}\,\mathrm{e}^{\pm\mathrm{i}kr}\right) = \pm\frac{k}{r^2}\frac{\vec{r}}{r}\,, \qquad (3.4.28)$$

which is radially *outgoing* for the upper sign, and *incoming* for the lower sign. Clearly, then, the upper sign is right for the situation depicted on page 102, and this is the choice to be made in (3.4.7).

In summary, now writing simply $|\mathrm{out}\rangle$ for $|\mathrm{out}_+\rangle$ and $\psi(\vec{r})$ for $\psi_+(\vec{r})$, we have the ket equation

$$|\mathrm{out}\rangle = |\mathrm{in}\rangle + (E - H_0 + \mathrm{i}\epsilon)^{-1}V|\mathrm{out}\rangle \qquad (3.4.29)$$

and the wave-function equation

$$\psi(\vec{r}) = \phi(\vec{r}) - \frac{M}{2\pi\hbar^2}\int\left(\mathrm{d}\vec{r}'\right)\frac{\mathrm{e}^{\mathrm{i}k|\vec{r}-\vec{r}'|}}{|\vec{r}-\vec{r}'|}V(\vec{r}')\psi(\vec{r}') \qquad (3.4.30)$$

for $V = V(\vec{R})$ or, after undoing the step from (3.4.9) to (3.4.11),

$$\psi(\vec{r}) = \phi(\vec{r}) - \frac{M}{2\pi\hbar^2}\int\left(\mathrm{d}\vec{r}'\right)\frac{\mathrm{e}^{\mathrm{i}k|\vec{r}-\vec{r}'|}}{|\vec{r}-\vec{r}'|}\langle\vec{r}'|V|\mathrm{out}\rangle \qquad (3.4.31)$$

for more general scattering potentials. These equations are of great importance in scattering theory, and all three are known as the *Lippmann–Schwinger equation*.

We also have the asymptotic form that applies far away from the scattering region,

$$\psi(\vec{r}) = \phi(\vec{r}) - \frac{M}{2\pi\hbar^2}\frac{\mathrm{e}^{\mathrm{i}kr}}{r}\int\left(\mathrm{d}\vec{r}'\right)\,\mathrm{e}^{-\mathrm{i}\vec{k}'\cdot\vec{r}'}\langle\vec{r}'|V|\mathrm{out}\rangle \qquad (3.4.32)$$

where

$$\vec{k}' = k\frac{\vec{r}}{r} \qquad (3.4.33)$$

is the direction of the outgoing current, expressed as a wave vector:

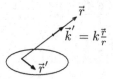

We take a plane wave propagating with wave vector $\vec{k}$ for the incoming amplitude,

$$\phi(\vec{r}) = \langle \vec{r} | \vec{k} \rangle = \frac{1}{(2\pi)^{3/2}} e^{i\vec{k} \cdot \vec{r}} \tag{3.4.34}$$

and then have the asymptotic form

$$\psi(\vec{r}) = \frac{1}{(2\pi)^{3/2}} \left[\underbrace{e^{i\vec{k} \cdot \vec{r}}}_{} + \underbrace{\frac{e^{ikr}}{r} f(\vec{k}', \vec{k})}_{} \right] \tag{3.4.35}$$

$$\begin{array}{cc} \text{incoming} & \text{outgoing, scattered} \\ \text{plane wave} & \text{spherical wave} \end{array}$$

with the *scattering amplitude*

$$f(\vec{k}', \vec{k}) = -(2\pi)^{3/2} \frac{M}{2\pi\hbar^2} \int (\mathrm{d}\vec{r}') \underbrace{e^{-i\vec{k}' \cdot \vec{r}'}}_{= (2\pi)^{3/2} \langle \vec{k}' | \vec{r}' \rangle} \langle \vec{r}' | V | \mathrm{out} \rangle$$

$$= -\left(\frac{2\pi}{\hbar} \right)^2 M \int (\mathrm{d}\vec{r}') \, \langle \vec{k}' | \vec{r}' \rangle \langle \vec{r}' | V | \mathrm{out} \rangle \tag{3.4.36}$$

or

$$f(\vec{k}', \vec{k}) = -\left(\frac{2\pi}{\hbar} \right)^2 M \langle \vec{k}' | V | \mathrm{out} \rangle. \tag{3.4.37}$$

The $\vec{k}$ dependence of $f(\vec{k}', \vec{k})$ is implicit, and complicated, inasmuch as it is contained in $|\mathrm{out}\rangle$ which is the solution of the Lippmann–Schwinger equation (3.4.29) for $|\mathrm{in}\rangle = |\vec{k}\rangle$.

The probability current of the outgoing spherical wave is

$$
\vec{\jmath}_{\text{out}} = \frac{\hbar}{M} \text{Im} \left(\frac{1}{(2\pi)^{3/2}} \frac{e^{-ikr}}{r} f(\vec{k}',\vec{k})^* \vec{\nabla} \frac{1}{(2\pi)^{3/2}} \frac{e^{ikr}}{r} f(\vec{k}',\vec{k}) \right)
$$

$$
= \frac{\hbar}{M} \frac{1}{(2\pi)^3} \frac{1}{r^2} \left[\left| f(\vec{k}',\vec{k}) \right|^2 \text{Im} \left(r\,e^{-ikr} \vec{\nabla} \frac{e^{ikr}}{r} \right) \right.
$$

$$
\left. + \text{Im} \left(f(\vec{k}',\vec{k})^* \vec{\nabla} f(\vec{k}',\vec{k}) \right) \right]
$$

$$
= \frac{\hbar}{M} \frac{1}{(2\pi)^3} \frac{1}{r^2} \left| f(\vec{k}',\vec{k}) \right|^2 k \frac{\vec{r}}{r} + \cdots , \tag{3.4.38}
$$

where the ellipsis stands for the terms that result from

$$
\vec{\nabla} f(\vec{k}',\vec{k}) = \vec{\nabla} f\left(k\frac{\vec{r}}{r},\vec{k}\right). \tag{3.4.39}
$$

Since $\frac{\vec{r}}{r}$ depends only on the direction of $\vec{r}$ but not on its length r, these terms are orthogonal to the radial direction $\frac{\vec{r}}{r}$ and do not contribute to the outgoing probability flux.

3-13 Verify that

$$
\vec{\nabla} e^{-i\vec{k}' \cdot \vec{r}'} = -i \frac{k}{r^3} \vec{r} \times (\vec{r}' \times \vec{r})\, e^{-i\vec{k}' \cdot \vec{r}'}
$$

and then state the missing terms of (3.4.38) explicitly.

This outgoing flux is the flux through solid angle $d\Omega$ in the direction of $\frac{\vec{r}}{r}$, with the surface element $d\vec{S} = \frac{\vec{r}}{r} r^2 d\Omega$. This outgoing flux is therefore given by

$$
\vec{\jmath}_{\text{out}} \cdot \frac{\vec{r}}{r} r^2 d\Omega = \frac{\hbar}{M} \frac{k}{(2\pi)^3} \left| f(\vec{k}',\vec{k}) \right|^2 d\Omega . \tag{3.4.40}
$$

We compare it with the incoming flux per unit area, that is: the incoming probability current density,

$$
\vec{\jmath}_{\text{in}} = \frac{\hbar}{M} \text{Im} \left(\frac{1}{(2\pi)^{3/2}} e^{-i\vec{k} \cdot \vec{r}} \vec{\nabla} \frac{1}{(2\pi)^{3/2}} e^{i\vec{k} \cdot \vec{r}} \right)
$$

$$
= \frac{\hbar}{M} \frac{1}{(2\pi)^3} \vec{k} , \tag{3.4.41}
$$

and identify the differential scattering cross section in accordance with

$$\vec{\jmath}_{\text{out}} \cdot r^2 \frac{\vec{r}}{r} d\Omega = \left| \vec{\jmath}_{\text{in}} \right| \frac{d\sigma}{d\Omega} d\Omega \,, \tag{3.4.42}$$

with the outcome

$$\frac{d\sigma}{d\Omega} = \left| f(\vec{k}', \vec{k}) \right|^2 . \tag{3.4.43}$$

This important relation between the differential cross section and the scattering amplitude justifies the name chosen for $f(\vec{k}', \vec{k})$: you get the scattering probability by squaring the scattering amplitude.

3.4.1 *Born approximation*

Whenever it is physically reasonable to expect that the effect of the scattering potential is not dominating, such as when the energy of the incoming particles is large, $|\text{out}\rangle \cong |\text{in}\rangle$ should be a good approximation. In any case $|\text{out}\rangle - |\text{in}\rangle$ is of first order in V (to lowest order) and so is, therefore, $f(\vec{k}', \vec{k})$. Accordingly, in lowest order in V we get $f(\vec{k}', \vec{k})$, and then $\frac{d\sigma}{d\Omega}$, by putting $|\text{out}\rangle \cong |\text{in}\rangle$. This approximation gives

$$f(\vec{k}', \vec{k}) = -\left(\frac{2\pi}{\hbar} \right)^2 M \langle \vec{k}' | V | \vec{k} \rangle \tag{3.4.44}$$

in general and

$$f(\vec{k}', \vec{k}) = -\left(\frac{2\pi}{\hbar} \right)^2 M \int \frac{(d\vec{r})}{(2\pi)^3} \, e^{i(\vec{k} - \vec{k}') \cdot \vec{r}} V(\vec{r})$$

$$= -\frac{M}{2\pi\hbar^2} \int (d\vec{r}) \, e^{i(\vec{k} - \vec{k}') \cdot \vec{r}} V(\vec{r}) \tag{3.4.45}$$

if V depends on $\vec{R}$ only. In the latter case, we thus get

$$\frac{d\sigma}{d\Omega} = \left| \frac{M}{2\pi\hbar^2} \int (d\vec{r}) \, e^{i(\vec{k} - \vec{k}') \cdot \vec{r}} V(\vec{r}) \right|^2 , \tag{3.4.46}$$

which is exactly the golden-rule result of (3.3.14). In the context of the Lippmann–Schwinger equation one speaks of the (first-order or lowest-order) *Born approximation*. As we shall see shortly, it is the leading term in a systematic expansion.

3-14   Find the Born approximation for $\dfrac{d\sigma}{d\Omega}$ for the "hard-sphere potential"

$$V(\vec{r}) = \begin{cases} V_0 & \text{if} \quad r = |\vec{r}| < a, \\ 0 & \text{if} \quad r > a. \end{cases}$$

Compare the total cross section with πa^2.

3-15   Find the Born approximation for $\dfrac{d\sigma}{d\Omega}$ for the gaussian potential

$$V(\vec{r}) = V_0 \, e^{-\frac{1}{2}(r/a)^2} \quad \text{with} \quad a > 0 \quad \text{and} \quad V_0 = \frac{(\hbar/a)^2}{2M}.$$

Then determine the total cross section σ in terms of E/V_0. What is the dominating E dependence for $E \ll V_0$ and $E \gg V_0$?

3.4.2   *Transition operator*

When it comes to determining the cross section, $|\text{out}\rangle$ itself does not matter as much as $V|\text{out}\rangle$, and the mapping $|\text{in}\rangle \rightarrow V|\text{out}\rangle$ is more relevant than the mapping $|\text{in}\rangle \rightarrow |\text{out}\rangle$. One introduces, therefore, the so-called *transition operator* T by

$$|\text{in}\rangle \rightarrow V|\text{out}\rangle = T|\text{in}\rangle. \tag{3.4.47}$$

With $|\text{in}\rangle = |\vec{k}\rangle$, the scattering amplitude (3.4.37) is

$$f(\vec{k}', \vec{k}) = -\left(\frac{2\pi}{\hbar}\right)^2 M\langle \vec{k}'|T|\vec{k}\rangle, \tag{3.4.48}$$

where the essential ingredient is the matrix element of T for the ket $|\vec{k}\rangle$ of the incoming plane wave with wave vector $\vec{k}$ and the bra $\langle \vec{k}'|$ of the outgoing plane wave with wave vector $\vec{k}'$.

On the search for an equation that determines the transition operator directly, that is: without the need of first finding $|\text{out}\rangle$, we return to the ket version (3.4.29) of the Lippmann–Schwinger equation,

$$|\text{out}\rangle = |\text{in}\rangle + (E - H_0 + i\epsilon)^{-1}V|\text{out}\rangle, \tag{3.4.49}$$

and note that it implies

$$\begin{aligned} T|\text{in}\rangle = V|\text{out}\rangle &= V|\text{in}\rangle + V(E - H_0 + i\epsilon)^{-1}T|\text{in}\rangle \\ &= \left[V + V(E - H_0 + i\epsilon)^{-1}T\right]|\text{in}\rangle. \end{aligned} \tag{3.4.50}$$

Since this must be true for all "in" states $|\text{in}\rangle$ (of energy E), we infer that

$$T = V + V(E - H_0 + i\epsilon)^{-1}T \qquad (3.4.51)$$

must hold.

This equation has a structure we have seen before, notably in (2.2.10). As we did there, here also we generate a systematic sequence of approximations by the iteration

$$T_{n+1} = V + V(E - H_0 + i\epsilon)^{-1}T_n, \qquad (3.4.52)$$

starting with $T_1 = V$, which is the analog of (2.2.11). Writing

$$G = (E - H_0 + i\epsilon)^{-1} \qquad (3.4.53)$$

for the Green's operator with H_0, we have

$$
\begin{aligned}
T_2 &= V + VGV, \\
T_3 &= V + VGV + VGVGV, \\
T_4 &= V + VGV + VGVGV + VGVGVGV,
\end{aligned}
\qquad (3.4.54)
$$

and so forth. Thinking of this graphically, we note that

$T_1:$ $\quad\overset{\vec{k}}{\longrightarrow}\underset{V}{\bullet}\overset{\vec{k}'}{\nearrow}$ $\qquad$ single scattering

$T_2:$ $\quad T_1 + \overset{\vec{k}}{\longrightarrow}\underset{V}{\bullet}\underset{V}{\overset{G}{\longrightarrow}}\bullet\overset{\vec{k}'}{\nearrow}$ $\qquad$ add double scattering

$T_3:$ $\quad T_2 + \overset{\vec{k}}{\longrightarrow}\underset{V}{\bullet}\overset{G}{\longrightarrow}\underset{V}{\bullet}\overset{G}{\searrow}\underset{V}{\bullet}\overset{\vec{k}'}{\nearrow}$ $\qquad$ add triple scattering

and so forth.

What we get is the series

$$T = \sum_{k=1}^{\infty} V(GV)^{k-1} \qquad (3.4.55)$$

and its truncations

$$T_n = \sum_{k=1}^{n} V(GV)^{k-1} = V + VGV + \cdots + \underbrace{VGV\cdots VGV}_{n \text{ times } V} \qquad (3.4.56)$$

which are known as *the* Born series and its nth-order approximation. For $n = 1$, we have $T_1 = V$, the 1st-order approximation with which the recursion (3.4.52) starts. Using $T \cong T_1 = V$ is, of course, just the Born approximation of Section 3.4.1.

3-16 Show that

$$T = V\frac{1}{1 - GV} = V\frac{1}{E - H + \mathrm{i}\epsilon}(E - H_0)$$

and explain why these explicit expression are not so useful in practice.

3.4.3 *Optical theorem*

For $\vec{k}' = \vec{k}$, we have *forward scattering*, that is the scattered spherical wave is aligned with the incoming plane wave, so that they can and will interfere. In particular, if there is a rather solid object in the way, we get a shadow of no or low intensity, which is to say that the interference must be destructive. More generally, the interference will lead to a *reduction* of the total forward flux, because what is scattered aside can no longer continue to propagate in the forward direction. Not surprisingly, then, there is a relation between $f(\vec{k}', \vec{k})$, the forward scattering amplitude, and the total cross section σ. This relation is the *optical theorem* stated in (3.4.69) below.

To derive it, we use the definition of the transition operator in (3.4.47) for $|\mathrm{in}\rangle = |\vec{k}\rangle$,

$$T|\vec{k}\rangle = V|\mathrm{out}\rangle \tag{3.4.57}$$

and its adjoint

$$\langle\vec{k}|T^\dagger = \langle\mathrm{out}|V\,, \tag{3.4.58}$$

to establish first

$$f(\vec{k}, \vec{k})^* = -\left(\frac{2\pi}{\hbar}\right)^2 M\langle\vec{k}|T|\vec{k}\rangle^* = -\left(\frac{2\pi}{\hbar}\right)^2 M\langle\vec{k}|T^\dagger|\vec{k}\rangle$$

$$= -\left(\frac{2\pi}{\hbar}\right)^2 M\langle\mathrm{out}|V|\vec{k}\rangle\,. \tag{3.4.59}$$

The ket version (3.4.29) of the Lippmann–Schwinger equation tells us that

$$|\vec{k}\rangle = |\mathrm{in}\rangle = |\mathrm{out}\rangle - (E - H_0 + \mathrm{i}\epsilon)^{-1}V|\mathrm{out}\rangle\,, \tag{3.4.60}$$

so that

$$f(\vec{k},\vec{k})^* = -\left(\frac{2\pi}{\hbar}\right)^2 M\left[\langle\text{out}|V|\text{out}\rangle - \langle\text{out}|V\frac{1}{E - H_0 + i\epsilon}V|\text{out}\rangle\right],$$
(3.4.61)

or with (3.4.57) and (3.4.58)

$$f(\vec{k},\vec{k})^* = -\left(\frac{2\pi}{\hbar}\right)^2 M\left[\langle\text{out}|V|\text{out}\rangle - \langle\vec{k}|T^\dagger\frac{1}{E - H_0 + i\epsilon}T|\vec{k}\rangle\right].$$
(3.4.62)

We proceed by utilizing the identity (2.5.56) for $x = E - H_0$,

$$\left.\frac{1}{E - H_0 + i\epsilon}\right|_{\epsilon \to 0} = \mathcal{P}\frac{1}{E - H_0} - i\pi\delta(E - H_0),$$
(3.4.63)

and extract the imaginary part,

$$\begin{aligned}
\text{Im}\left(f(\vec{k},\vec{k})\right) &= -\text{Im}\left(f(\vec{k},\vec{k})^*\right) \\
&= \left(\frac{2\pi}{\hbar}\right)^2 M\pi\langle\vec{k}|T^\dagger\delta(E - H_0)T|\vec{k}\rangle \\
&= \left(\frac{2\pi}{\hbar}\right)^2 M\pi\int\left(d\vec{k'}\right)\langle\vec{k}|T^\dagger|\vec{k'}\rangle\,\delta\left(E - \frac{(\hbar k')^2}{2M}\right)\langle\vec{k'}|T|\vec{k}\rangle
\end{aligned}$$
(3.4.64)

where we have used the completeness of the $|\vec{k'}\rangle$ states and the eigenvalue equation $H_0|\vec{k'}\rangle = |\vec{k'}\rangle(\hbar k')^2/(2M)$.

Now, recalling the fundamental link (3.4.48) between the transition operator and the scattering amplitude,

$$\langle\vec{k'}|T|\vec{k}\rangle = -\left(\frac{\hbar}{2\pi}\right)^2\frac{1}{M}f(\vec{k'},\vec{k}),$$
(3.4.65)

we note that

$$\begin{aligned}
\langle\vec{k}|T^\dagger|\vec{k'}\rangle\langle\vec{k'}|T|\vec{k}\rangle &= \left|\langle\vec{k'}|T|\vec{k}\rangle\right|^2 \\
&= \left(\frac{\hbar}{2\pi}\right)^4\frac{1}{M^2}\left|f(\vec{k'},\vec{k})\right|^2 \\
&= \left(\frac{\hbar}{2\pi}\right)^4\frac{1}{M^2}\frac{d\sigma}{d\Omega}.
\end{aligned}$$
(3.4.66)

And upon using (3.3.11) once more,

$$\delta\left(E - \frac{(\hbar k')^2}{2M}\right) = \delta\left(\frac{(\hbar k)^2}{2M} - \frac{(\hbar k')^2}{2M}\right) = \frac{M}{\hbar^2 k}\delta(k - k'), \qquad (3.4.67)$$

we get

$$\begin{aligned}
\operatorname{Im}\left(f(\vec{k}, \vec{k})\right) &= \frac{1}{4\pi k} \int \left(\mathrm{d}\vec{k}'\right) \delta(k - k')\frac{\mathrm{d}\sigma}{\mathrm{d}\Omega} \\
&= \frac{1}{4\pi k} \underbrace{\int_0^\infty \mathrm{d}k'\, k'^2 \delta(k - k')}_{= k^2} \underbrace{\int \mathrm{d}\Omega\, \frac{\mathrm{d}\sigma}{\mathrm{d}\Omega}}_{= \sigma} \\
&= \frac{k}{4\pi}\sigma \qquad\qquad\qquad\qquad\qquad\qquad (3.4.68)
\end{aligned}$$

and finally arrive at

$$\sigma = \frac{4\pi}{k}\operatorname{Im}\left(f(\vec{k}', \vec{k})\right) . \qquad (3.4.69)$$

This is the famous *optical theorem*, first established by Niels H. D. Bohr, Rudolf Peierls, and George Placzek, but as the name indicates the theorem has a historical precursor and analog in classical optics.

3-17 For

$$\psi(\vec{r}, t) = \frac{1}{(2\pi)^{3/2}}\left[\mathrm{e}^{\mathrm{i}\vec{k}\cdot\vec{r}} + \frac{\mathrm{e}^{\mathrm{i}kr}}{r}f(\theta)\right]\mathrm{e}^{-\mathrm{i}\hbar k^2 t/(2M)},$$

where θ is the deflection angle, $\vec{k}\cdot\vec{k}' = k^2\cos\theta$, determine the probability current $\vec{\jmath}(\vec{r}, t)$ and verify the optical theorem directly by a calculation of the total flux through a sphere of very large radius. Hint: First show that the total flux vanishes, then examine the various terms that contribute to it, and remember that this $\psi(\vec{r}, t)$ represents a large-r approximation.

3.4.4 *Example of an exact solution*

There are very few examples for which the Lippmann–Schwinger equation has an exact analytical solution. One such example is the separable potential of Exercise 3-9 on page 99,

$$V = |s\rangle E_0 \langle s| \qquad (3.4.70)$$

with $\langle s|s\rangle = 1$.

Here we have

$$|\text{out}\rangle = |\text{in}\rangle + GV|\text{out}\rangle = |\text{in}\rangle + G|s\rangle E_0\langle s|\text{out}\rangle, \qquad (3.4.71)$$

and we first determine $\langle s|\text{out}\rangle$ from

$$\langle s|\text{out}\rangle = \langle s|\text{in}\rangle + \underbrace{\langle s|G|s\rangle E_0}_{=\,\text{tr}\{GV\}}\langle s|\text{out}\rangle, \qquad (3.4.72)$$

namely

$$\langle s|\text{out}\rangle = \frac{\langle s|\text{in}\rangle}{1 - \text{tr}\{GV\}}. \qquad (3.4.73)$$

The ket $|\text{out}\rangle$ itself is then known,

$$\begin{aligned} |\text{out}\rangle &= |\text{in}\rangle + G|s\rangle E_0 \frac{\langle s|\text{in}\rangle}{1 - \text{tr}\{GV\}} \\ &= |\text{in}\rangle + \frac{GV|\text{in}\rangle}{1 - \text{tr}\{GV\}}, \end{aligned} \qquad (3.4.74)$$

and we get the transition operator from its definition

$$T|\text{in}\rangle = V|\text{out}\rangle = \left(V + \frac{VGV}{1 - \text{tr}\{GV\}}\right)|\text{in}\rangle, \qquad (3.4.75)$$

that is

$$T = V + \frac{VGV}{1 - \text{tr}\{GV\}}. \qquad (3.4.76)$$

Since $V = |s\rangle E_0\langle s|$ here, we have $VGV = V\,\text{tr}\{GV\}$ and

$$T = \frac{V}{1 - \text{tr}\{GV\}} \qquad (3.4.77)$$

is the final compact expression.

3-18 Derive this also from (3.4.51), the equation obeyed by T,

$$T = V + VGT.$$

3-19 The comparison of the particular result (3.4.77) with the general expression for T in Exercise 3-16 on page 114 shows that we have $GV \rightarrow \text{tr}\{GV\}$ effectively. Explain why this replacement is justified.

3-20 For $|s\rangle$ such that

$$\langle \vec{k}|s\rangle = \frac{2}{\pi}\sqrt{2k_0^5}\,\frac{1}{\left(k^2 + k_0^2\right)^2} \quad \text{with} \quad k_0 > 0\,,$$

evaluate $\mathrm{tr}\{GV\}$ (a bit tough!), and then find $\dfrac{\mathrm{d}\sigma}{\mathrm{d}\Omega}$ and σ. Then check that the optical theorem holds.

3.5   Partial waves

If the scattering potential is spherically symmetric, $V(\vec{r}) = V(r)$, then the Hamilton operator $H = \vec{P}^2/(2M) + V(R)$ commutes with the angular momentum vector operator $\vec{L} = \vec{R} \times \vec{P}$,

$$\left[H, \vec{L}\right] = 0\,. \tag{3.5.1}$$

Therefore, recalling a lesson in Section 4.4 of *Simple Systems*, we can have a basis set of states that are common eigenstates of H, $\vec{L}^2$, and L_3,

$$\begin{aligned}
H|E,l,m\rangle &= |E,l,m\rangle E\,,\\
\vec{L}^2|E,l,m\rangle &= |E,l,m\rangle \hbar^2 l(l+1)\,,\\
L_z|E,l,m\rangle &= |E,l,m\rangle \hbar m\,,
\end{aligned} \tag{3.5.2}$$

with $m = 0, \pm 1, \pm 2, \ldots, \pm l$ for the given l value, which itself is an integer, $l = 0, 1, 2, \ldots$, and the energy E can, and usually will, depend on the quantum number l but not on m. We further recall from *Simple Systems* that the position wave function $\langle \vec{r}|E,l,m\rangle$ is conveniently written as

$$\langle \vec{r}|E,l,m\rangle = \frac{1}{r}u_l(r)Y_{lm}(\vartheta,\varphi) \tag{3.5.3}$$

where r, ϑ, φ are the usual spherical coordinates of $\vec{r}$ and $Y_{lm}(\vartheta,\varphi)$ is one of the spherical harmonics. The radial function $u_l(r)$ obeys the radial Schrödinger equation

$$\left[-\frac{\hbar^2}{2M}\frac{\partial^2}{\partial r^2} + \frac{\hbar^2 l(l+1)}{2Mr^2} + V(r)\right]u_l(r) = Eu_l(r) \tag{3.5.4}$$

where the "centrifugal potential" $\dfrac{\hbar^2 l(l+1)}{2Mr^2}$ is added to the physical potential energy $V(r)$.

We apply these matters to the scattering situation. As a consequence of the spherical symmetry, all directions in space are equivalent, and so we can agree to have the incoming plane wave propagating in the z direction, $\vec{k} = k\vec{e}_z$ with $k > 0$. Then

$$\phi(r) = \langle \vec{r} | \vec{k} \rangle = \frac{1}{(2\pi)^{3/2}}\,\mathrm{e}^{\mathrm{i}kz} = \frac{1}{(2\pi)^{3/2}}\,\mathrm{e}^{\mathrm{i}kr\cos\theta}\,, \qquad (3.5.5)$$

recognizing that the polar angle ϑ of spherical coordinates is actually the deflection angle θ:

Therefore, the expansion of $\phi(\vec{r})$ into a sum of the form $\displaystyle\sum_{lm}\frac{1}{r}u_l(r)Y_{lm}(\vartheta,\varphi)$ will only involve the φ-independent $m = 0$ terms, for which $Y_{l0}(\vartheta,\varphi) \propto P_l(\cos\vartheta)$ is essentially the lth Legendre polynomial of $\cos\vartheta$ (Adrien M. Legendre). In short, we have the *partial-wave expansion*

$$\mathrm{e}^{\mathrm{i}kr\cos\theta} = \sum_{l=0}^{\infty}(2l+1)\mathrm{i}^l j_l(kr)P_l(\cos\theta) \qquad (3.5.6)$$

of the plane wave $\mathrm{e}^{\mathrm{i}kz}$, whereby $(2l+1)\mathrm{i}^l j_l(kr)$ is a convenient way of writing what plays the role of $\frac{1}{r}u_l(r)$ here. The functions $j_l(kr)$ thus defined are known as the *spherical Bessel functions*, named after Friedrich W. Bessel. We state them explicitly by invoking the orthonormality relation of the Legendre polynomials,

$$\frac{1}{2}\int_{-1}^{1}\mathrm{d}\zeta\, P_l(\zeta)P_{l'}(\zeta) = \frac{\delta_{ll'}}{2l+1}\,, \qquad (3.5.7)$$

so that

$$\mathrm{i}^l j_l(kr) = \frac{1}{2}\int_{-1}^{1}\mathrm{d}\zeta\, P_l(\zeta)\,\mathrm{e}^{\mathrm{i}kr\zeta}\,. \qquad (3.5.8)$$

3-21 Use $P_l(-\zeta) = (-1)^l P_l(\zeta)$ to show that $j_l(kr)$ is real. This is the reason for the i^l prefactor in (3.5.6).

The differential equation obeyed by $rj_l(kr)$ is (3.5.4), the radial Schrödinger equation for $u_l(r)$, with $V(r) = 0$ and $E = (\hbar k)^2/(2M)$, that is

$$\left[\frac{\partial^2}{\partial r^2} - \frac{l(l+1)}{r^2} + k^2\right] rj_l(kr) = 0 \,. \tag{3.5.9}$$

For the present application to scattering we do not need to know the $j_l(kr)$s in detail, but we must understand their asymptotic form for $kr \gg 1$. This is available by an integration by parts,

$$\mathrm{i}^l j_l(kr) = \frac{1}{2} \int_{-1}^{1} \mathrm{d}\zeta \, P_l(\zeta) \frac{\partial}{\partial \zeta} \frac{\mathrm{e}^{\mathrm{i}kr\zeta}}{\mathrm{i}kr}$$

$$= \underbrace{\frac{1}{2} P_l(\zeta) \frac{\mathrm{e}^{\mathrm{i}kr\zeta}}{\mathrm{i}kr}\bigg|_{\zeta=-1}^{+1}}_{\propto 1/(kr)} - \underbrace{\frac{1}{2} \int_{-1}^{1} \mathrm{d}\zeta \, \frac{\partial P_l(\zeta)}{\partial \zeta} \frac{\mathrm{e}^{\mathrm{i}kr\zeta}}{\mathrm{i}kr}}_{\substack{\text{terms} \propto (kr)^{-2}, (kr)^{-3}, \dots \\ \text{obtained by successive} \\ \text{integrations by part}}} \tag{3.5.10}$$

so that the leading asymptotic terms are

$$\mathrm{i}^l j_l(kr) \cong \frac{1}{2} \underbrace{P_l(+1)}_{=1} \frac{\mathrm{e}^{\mathrm{i}kr}}{\mathrm{i}kr} - \frac{1}{2} \underbrace{P_l(-1)}_{=(-1)^l = \mathrm{i}^{2l}} \frac{\mathrm{e}^{-\mathrm{i}kr}}{\mathrm{i}kr} \,, \tag{3.5.11}$$

and we have

$$j_l(kr) \cong \frac{1}{2\mathrm{i}kr}\left(\mathrm{i}^{-l}\,\mathrm{e}^{\mathrm{i}kr} - \mathrm{i}^l\,\mathrm{e}^{-\mathrm{i}kr}\right)$$

$$= \frac{1}{2\mathrm{i}kr}\left(\mathrm{e}^{\mathrm{i}kr - \mathrm{i}l\frac{\pi}{2}} - \mathrm{e}^{-\mathrm{i}kr + \mathrm{i}l\frac{\pi}{2}}\right)$$

$$= \frac{1}{kr}\sin\left(kr - l\frac{\pi}{2}\right) \tag{3.5.12}$$

for $kr \gg 1$.

3-22 Find $j_0(kr)$, $j_1(kr)$, and $j_2(kr)$ explicitly and verify this large-kr behavior for them.

We note that

$$\phi(\vec{r}) \cong \frac{1}{(2\pi)^{3/2}} \sum_{l=0}^{\infty} (2l+1) i^l \frac{1}{kr} \sin\left(kr - l\frac{\pi}{2}\right) P_l(\cos\theta)$$

$$= \frac{1}{(2\pi)^{3/2}} \sum_{l=0}^{\infty} (2l+1) \left(\frac{e^{ikr}}{2ikr} - (-1)^l \frac{e^{-ikr}}{2ikr}\right) P_l(\cos\theta) \quad (3.5.13)$$

consists of a superposition of outgoing spherical waves $\propto e^{ikr}/r$ and incoming spherical waves $\propto e^{-ikr}/r$. The total wave function

$$\psi(\vec{r}) = \phi(\vec{r}) + \frac{1}{(2\pi)^{3/2}} \frac{e^{ikr}}{r} f(\vec{k}', \vec{k}) \quad (3.5.14)$$

has an additional outgoing spherical wave, the amplitude of the scattered wave. As a consequence, the $\psi_l(kr)$s in the partial-wave expansion

$$\psi(\vec{r}) = \frac{1}{(2\pi)^{3/2}} \sum_{l} (2l+1) i^l \psi_l(kr) P_l(\cos\theta) , \quad (3.5.15)$$

which obey the differential equation (3.5.4) for $u_l(r) = r\psi_l(r)$,

$$\left[\frac{\partial^2}{\partial r^2} - \frac{l(l+1)}{r^2} - \frac{2M}{\hbar^2} V(r) + k^2\right] r\psi_l(kr) = 0 \quad (3.5.16)$$

for all r, and the $V = 0$ equation (3.5.9)

$$\left[\frac{\partial^2}{\partial r^2} - \frac{l(l+1)}{r^2} + k^2\right] r\psi_l(kr) = 0 \quad (3.5.17)$$

in the asymptotic region ($kr \gg 1$) of present interest, must be such that

$$kr \gg 1 : \quad kr\psi_l(kr) = e^{i\delta_l} \sin\left(kr - l\frac{\pi}{2} + \delta_l\right). \quad (3.5.18)$$

This is the only option allowed by the asymptotic form of the differential equation subject to the constraint that the incoming spherical wave in

$$i^l \psi_l(kr) \cong e^{i\delta_l} \frac{i^l}{kr} \sin\left(kr - l\frac{\pi}{2} + \delta_l\right)$$

$$= \frac{1}{2ikr} e^{i\delta_l} e^{\frac{i}{2}\pi l} \left(e^{ikr - \frac{i}{2}\pi l + i\delta_l} - e^{-ikr + \frac{i}{2}\pi l - i\delta_l}\right)$$

$$= \frac{1}{2ikr} \left(\underbrace{e^{ikr + 2i\delta_l}}_{\substack{\text{outgoing,} \\ \text{modified}}} - \underbrace{(-1)^l e^{-ikr}}_{\substack{\text{incoming,} \\ \text{as before}}}\right) \quad (3.5.19)$$

is not altered. Thus, the net effect of the scattering potential is to introduce the *scattering phases* δ_l in

$$j_l(kr) \cong \frac{1}{kr} \sin\left(kr - \frac{1}{2}\pi l\right) \longrightarrow \frac{e^{i\delta_l}}{kr} \sin\left(kr - \frac{1}{2}\pi l + \delta_l\right) \qquad (3.5.20)$$

and the scattering amplitude is then obtained as

$$f(\vec{k}', \vec{k}) = \frac{1}{k} \sum_{l=0}^{\infty} (2l+1) \frac{1}{2i} \left(e^{2i\delta_l} - 1\right) P_l(\cos\theta). \qquad (3.5.21)$$

Recognizing that $f(\vec{k}', \vec{k})$ depends on θ predominantly (and on k through the k dependence of the δ_l), we write this as

$$f(\theta) = \frac{1}{k} \sum_{l=0}^{\infty} (2l+1) e^{i\delta_l} \sin\delta_l \, P_l(\cos\theta), \qquad (3.5.22)$$

which is the partial-wave expansion for the scattering amplitude. The problem of determining the scattering amplitude $f(\vec{k}', \vec{k}) \equiv f(\theta)$ is thus reduced to the calculation of the scattering phases δ_l. It is worth emphasizing that this expression for $f(\theta)$ is exact, it does not involve any approximations. Approximations will, however, have to be introduced, as a rule, when calculating the δ_ls.

The total cross section is easily obtained with the aid of the optical theorem of (3.4.69),

$$\sigma = \frac{4\pi}{k} \mathrm{Im}\big(f(\theta = 0)\big), \qquad (3.5.23)$$

inasmuch as $\vec{k}' = \vec{k}$ means $\theta = 0$ here. We have $\mathrm{Im}\left(e^{i\delta_l} \sin\delta_l\right) = (\sin\delta_l)^2$ and $P_l(\cos\theta = 1) = 1$, so that

$$\sigma = \frac{4\pi}{k^2} \sum_{l=0}^{\infty} (2l+1)(\sin\delta_l)^2. \qquad (3.5.24)$$

3-23 Show that you get the same, indeed, by integrating $\dfrac{d\sigma}{d\Omega} = |f(\theta)|^2$ over the solid angle $d\Omega = 2\pi d\theta \sin\theta$.

3.6 *s*-wave scattering

At low energies, the centrifugal potential $\propto l(l+1)/r^2$ suppresses the wave function at short distances,

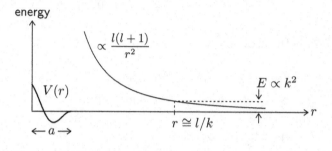

and then there is little or no effect of the scattering potential — except for the *s*-*waves* to $l = 0$, for which there is no centrifugal barrier. Low-energy scattering is, therefore, completely dominated by the $l = 0$ sector, inasmuch as $\delta_l \cong 0$ is a very good approximation for $l > 0$ when k is sufficiently small, more precisely: when

$$k \ll \frac{1}{a}, \quad a = \text{range of } V(r). \tag{3.6.1}$$

As an example of a short-range potential we consider the hard-sphere potential of Exercise 3-14 on page 112,

$$V(r) = \begin{cases} V_0 & \text{for } r < a, \\ 0 & \text{for } r > a, \end{cases} \tag{3.6.2}$$

with $V_0 > 0$ (repulsive potential) or $V_0 < 0$ (attractive potential). We have then

$$\left(\frac{\partial^2}{\partial r^2} + k^2 \right) u_0(r) = 0 \quad \text{for } r > a \tag{3.6.3}$$

and

$$\left(\frac{\partial^2}{\partial r^2} + k^2 - \frac{2MV_0}{\hbar^2} \right) u_0(r) = 0 \quad \text{for } r < a. \tag{3.6.4}$$

The $r > a$ part of the solution is the right-hand side of (3.5.18) for $l = 0$,

$$u_0(r) = e^{i\delta_0} \sin(kr + \delta_0) \quad \text{for } r > a. \tag{3.6.5}$$

For $r < a$, we distinguish

$$\text{(A)} \quad k^2 - \frac{2MV_0}{\hbar^2} = \kappa^2 > 0 \,,$$

$$\text{(B)} \quad k^2 - \frac{2MV_0}{\hbar^2} = -\kappa^2 < 0 \,, \tag{3.6.6}$$

with $\kappa > 0$ in both cases, and write

$$\text{(A)} \quad u_0(r) = \mathrm{e}^{\mathrm{i}\delta_0} C \sin(\kappa r) \,,$$

$$\text{(B)} \quad u_0(r) = \mathrm{e}^{\mathrm{i}\delta_0} C \sinh(\kappa r) \,, \tag{3.6.7}$$

whereby we incorporate the boundary condition $u_0(r = 0) = 0$.

The two unknowns — the phase shift δ_0 and the amplitude factor C — are determined by the two equations that state the continuity of $u_0(r)$ and $\frac{\mathrm{d}}{\mathrm{d}r} u_0(r)$ at $r = a$:

$$\begin{aligned}
\text{(A)} \quad & C \sin(\kappa a) = \sin(ka + \delta_0) \,, \\
& C\kappa \cos(\kappa a) = k \cos(ka + \delta_0) \,, \\
\text{(B)} \quad & C \sinh(\kappa a) = \sin(ka + \delta_0) \,, \\
& C\kappa \cosh(\kappa a) = k \cos(ka + \delta_0) \,. \tag{3.6.8}
\end{aligned}$$

We could extract rather explicit expression for δ_0, but do not need this much detail if we are mainly interested in the total cross section σ, for which $(\sin \delta_0)^2$ is needed.

3-24 Express δ_0 in terms of ka and κ/k. When do you get $\delta_0 > 0$, when $\delta_0 < 0$? What is the physical significance of the sign of δ_0?

We note that

$$\begin{aligned}
\sin \delta_0 &= \sin(ka + \delta_0 - ka) \\
&= \cos(ka) \sin(ka + \delta_0) - \sin(ka) \cos(ka + \delta_0) \,, \tag{3.6.9}
\end{aligned}$$

with the consequence

$$\sin \delta_0 = \begin{cases} C\left[\cos(ka) \sin(\kappa a) - \dfrac{\kappa}{k} \sin(ka) \cos(\kappa a) \right] & \text{for (A)} \,, \\[2mm] C\left[\cos(ka) \sinh(\kappa a) - \dfrac{\kappa}{k} \sin(ka) \cosh(\kappa a) \right] & \text{for (B)} \,. \end{cases} \tag{3.6.10}$$

With

$$\frac{1}{C^2} = \left[\frac{1}{C}\sin(ka+\delta_0)\right]^2 + \left[\frac{1}{C}\cos(ka+\delta_0)\right]^2$$

$$= \begin{cases} [\sin(\kappa a)]^2 + \left[\dfrac{\kappa}{k}\cos(\kappa a)\right]^2 & \text{for (A)}, \\[3mm] [\sinh(\kappa a)]^2 + \left[\dfrac{\kappa}{k}\cosh(\kappa a)\right]^2 & \text{for (B)}, \end{cases} \tag{3.6.11}$$

this gives

(A) $\quad (\sin\delta_0)^2 = \dfrac{[k\cos(ka)\sin(\kappa a) - \kappa\sin(ka)\cos(\kappa a)]^2}{[k\sin(\kappa a)]^2 + [\kappa\cos(\kappa a)]^2}$,

(B) $\quad (\sin\delta_0)^2 = \dfrac{[k\cos(ka)\sinh(\kappa a) - \kappa\sin(ka)\cosh(\kappa a)]^2}{[k\sinh(\kappa a)]^2 + [\kappa\cosh(\kappa a)]^2}$, $\tag{3.6.12}$

and the resulting cross section $\sigma = \left(4\pi/k^2\right)\left(\sin\delta_0\right)^2$ is

$$\sigma = 4\pi a^2 \times \begin{cases} \dfrac{\left[\cos(ka)\frac{\sin(\kappa a)}{\kappa a} - \frac{\sin(ka)}{ka}\cos(\kappa a)\right]^2}{\left[\frac{k}{\kappa}\sin(\kappa a)\right]^2 + [\cos(\kappa a)]^2} & \text{for (A)}, \\[6mm] \dfrac{\left[\cos(ka)\frac{\sinh(\kappa a)}{\kappa a} - \frac{\sin(ka)}{ka}\cosh(\kappa a)\right]^2}{\left[\frac{k}{\kappa}\sinh(\kappa a)\right]^2 + [\cosh(\kappa a)]^2} & \text{for (B)}. \end{cases} \tag{3.6.13}$$

This cross section applies to low-energy scattering by the hard-sphere potential (3.6.2) whereby (3.6.6) identifies the two cases.

It is interesting to consider the long-wavelength limit of $k \to 0$. Then we have (A) for an attractive potential, $V_0 < 0$, and (B) for a repulsive potential, $V_0 > 0$; and we get

$$\sigma = 4\pi a^2 \times \begin{cases} \left(\dfrac{\tan(\kappa a)}{\kappa a} - 1\right)^2 & \text{for } V_0 < 0, \\[4mm] \left(1 - \dfrac{\tanh(\kappa a)}{\kappa a}\right)^2 & \text{for } V_0 > 0, \end{cases} \tag{3.6.14}$$

and both give $\sigma = 0$ for $V_0 \to 0$. The argument of the tangent, or hyperbolic tangent, is

$$\kappa a = \sqrt{2M|V_0|a^2/\hbar^2} \tag{3.6.15}$$

in both cases.

For $V_0 > 0$,

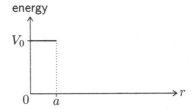

the limit $V_0 \to \infty$ is that of an impenetrable sphere, for which $u_0(r) = 0$ when $r < a$. This limit has

$$\frac{\tanh(\kappa a)}{\kappa a} \to 0, \quad \sigma \to 4\pi a^2, \qquad (3.6.16)$$

so that the cross section is four times the geometrical cross section πa^2 of the sphere with radius a. As one student remarked during lecture, σ happens to be the surface of the sphere, and this might be interpreted as the s-wave probing the sphere from all sides:

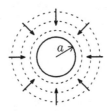

Indeed, this is a rather natural picture for the sphere-shaped wave fronts of an s-wave.

The situation is quite different for the attractive case $V_0 < 0$, where $|V_0| \to \infty$ means an ever stronger attractive potential with ever more bound states. Since $\tan(\kappa a) = \pm\infty$ when $\kappa a \to \frac{\pi}{2}, \frac{3\pi}{2}, \frac{5\pi}{2}, \ldots$, the cross section is very large for values of V_0 close to

$$V_0 = -\frac{\hbar^2}{2M}\left(\frac{\pi/2}{a}\right)^2, -\frac{\hbar^2}{2M}\left(\frac{3\pi/2}{a}\right)^2, -\frac{\hbar^2}{2M}\left(\frac{5\pi/2}{a}\right)^2, \ldots. \qquad (3.6.17)$$

These are the V_0 values for which the total number of bound states changes by one:

no bound state for
$$V_0 > -\frac{\hbar^2}{2M}\left(\frac{\pi}{2a}\right)^2,$$

one bound state for $-\dfrac{\hbar^2}{2M}\left(\dfrac{\pi}{2a}\right)^2 > V_0 > -\dfrac{\hbar^2}{2M}\left(\dfrac{3\pi}{2a}\right)^2,$

two bound states for $-\dfrac{\hbar^2}{2M}\left(\dfrac{3\pi}{2a}\right)^2 > V_0 > -\dfrac{\hbar^2}{2M}\left(\dfrac{5\pi}{2a}\right)^2,$ (3.6.18)

and so forth.

3-25 Justify this statement about the number of bound $l = 0$ states.

At one of those threshold values, there is a bound state, the new one, at energy $E = 0$, that is at the edge of the continuum of scattering states. This energetic degeneracy, or near degeneracy when V_0 is just a bit more negative, means that the *s*-waves with $k \gtrsim 0$ are nearly resonant with a bound state. Such a resonance can lead to a dramatic increase of the scattering cross section, as is illustrated by the example considered here.

3-26 A particle of mass M and kinetic energy $E = \dfrac{(\hbar k)^2}{2M}$ is scattered by the δ-shell potential

$$V(\vec{r}) = V_0 a\, \delta(r - a) \quad \text{with} \quad a > 0 \quad \text{and} \quad V_0 = \frac{(\hbar/a)^2}{2M}.$$

Determine the total cross section σ_0 for *s*-wave scattering. Write σ_0 in the form $\sigma_0 = \pi a^2 f(ka)$ with a suitable function $f(ka)$ of the product ka. What do you get for $ka \ll 1$?

Chapter 4

Angular Momentum

4.1 Spin

In *Basic Matters*, much of the general formalism of quantum mechanics is developed under the guidance of the example of magnetic silver atoms that pass through magnetic fields, are probed by inhomogeneous Stern–Gerlach magnets, and so forth. Wolfgang Pauli's vector operator $\vec{\sigma}$ is central to the description of a silver atom (in its magnetic properties), and we recall from (2.9.12) of *Basic Matters* that

$$\left[\vec{a} \cdot \vec{\sigma}, \vec{b} \cdot \vec{\sigma} \right] = 2\mathrm{i} \left(\vec{a} \times \vec{b} \right) \cdot \vec{\sigma} \tag{4.1.1}$$

is the commutation relation for two arbitrary components of $\vec{\sigma}$. The vector $\vec{\sigma}$ specifies the orientation of the magnetic moment carried by the atom, so that rotating the magnetic moment is tantamount to rotating the Pauli vector $\vec{\sigma}$. Such a rotation would be described by an equation of motion of the form

$$\frac{\mathrm{d}}{\mathrm{d}t}\vec{\sigma} = \vec{\omega} \times \vec{\sigma} = \frac{1}{\mathrm{i}\hbar}[\vec{\sigma}, H] , \tag{4.1.2}$$

where the relevant term in the Hamilton operator H is

$$H_{\mathrm{rot}} = \frac{\hbar}{2}\vec{\omega} \cdot \vec{\sigma} \quad \text{or} \quad H_{\mathrm{rot}} = \vec{\omega} \cdot \vec{S} \tag{4.1.3}$$

with

$$\vec{S} = \frac{\hbar}{2}\vec{\sigma} . \tag{4.1.4}$$

This $\vec{S}$ is thus the hermitian generator for internal rotations of the silver atom. It obeys the commutation relation

$$\left[\vec{a} \cdot \vec{S}, \vec{b} \cdot \vec{S}\right] = i\hbar\left(\vec{a} \times \vec{b}\right) \cdot \vec{S} \tag{4.1.5}$$

that we have seen in *Simple Systems* for the orbital angular momentum $\vec{L} = \vec{R} \times \vec{P}$,

$$\left[\vec{a} \cdot \vec{L}, \vec{b} \cdot \vec{L}\right] = i\hbar\left(\vec{a} \times \vec{b}\right) \cdot \vec{L}. \tag{4.1.6}$$

As we know from the discussion of $\vec{L}$, it is the hermitian generator for orbital rotation. For the silver atom, then, we need to distinguish between the *internal rotation*:

for a counter-clockwise rotation around an axis perpendicular to the paper, and *orbital rotation*:

during which the magnetic moment keeps pointing in the same direction. When both rotations happen at the same rate, we have the picture of a *rigid rotation*:

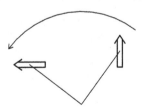

The respective terms in the Hamilton operator are

$$\vec{\omega} \cdot \vec{S} \quad \text{for the internal rotation only,}$$
$$\vec{\omega} \cdot \vec{L} \quad \text{for the orbital rotation only,}$$
$$\vec{\omega} \cdot \left(\vec{L} + \vec{S}\right) \quad \text{for both jointly.} \tag{4.1.7}$$

There is nothing so particular about the silver atom considered. We can, therefore, safely infer that this is just one example of the general situation that atomic objects have orbital angular momentum $\vec{L}$, internal angular momentum $\vec{S}$, commonly referred to as *spin*, and total angular momentum

$$\vec{J} = \vec{L} + \vec{S}. \tag{4.1.8}$$

For $\vec{J}$, the same commutation relations apply,

$$\left[\vec{a} \cdot \vec{J}, \vec{b} \cdot \vec{J}\right] = i\hbar\left(\vec{a} \times \vec{b}\right) \cdot \vec{J} \tag{4.1.9}$$

and it follows that $\vec{J}^2$ and J_z (or J_x, or J_y) have common eigenstates. Following the pattern that is used in (4.2.3) of *Simple Systems* for orbital angular momentum $\vec{L}$, we write $|j, m\rangle$ for the common eigenket, and state the eigenvalues by

$$J_z|j, m\rangle = |j, m\rangle \hbar m,$$
$$\vec{J}^2|j, m\rangle = |j, m\rangle \hbar^2 j(j + 1). \tag{4.1.10}$$

If there is orbital angular momentum only, that is $\vec{J} = \vec{L}$, we know that $j = 0, 1, 2, \ldots$ and $m = 0, \pm 1, \pm 2, \ldots, \pm j$ are the possible eigenvalues. By contrast, for the spin angular momentum of a silver atom, we have

$$\vec{J} = \frac{\hbar}{2}\vec{\sigma},$$
$$\vec{J}^2 = \left(\frac{\hbar}{2}\right)^2 (\sigma_x^2 + \sigma_y^2 + \sigma_z^2) = \frac{3}{4}\hbar^2$$
$$= \hbar^2 \frac{1}{2}\left(\frac{1}{2} + 1\right), \tag{4.1.11}$$

so that $j = \frac{1}{2}$ here, and the eigenvalues of $J_z = \frac{\hbar}{2}\sigma_z$ are $\pm\frac{\hbar}{2}$, that is $m = \pm\frac{1}{2}$. So, here we have half-integer values for j and m, whereas orbital angular momentum can have integer values only. Clearly, then, one cannot interpret the spin of a silver atom as coming about by the orbital motion of some constituents around a center inside the atom. Angular momentum of the spin type is of a quite different nature than orbital angular momentum. Spin is truly intrinsic to the atomic object.

We must, therefore, find out what are the eigenvalues for $\vec{J}^2$ and J_z in general. For this purpose we employ the same methodology as for orbital angular momentum in Section 4.1 of *Simple Systems*. In a first step, we introduce

$$J_\pm = J_x \pm \mathrm{i}J_y \tag{4.1.12}$$

and note that

$$
\begin{aligned}
[J_z, J_\pm] &= [J_z, J_x] \pm \mathrm{i}[J_z, J_y] \\
&= \mathrm{i}\hbar J_y \pm \mathrm{i}(-\mathrm{i}\hbar J_x) = \pm\hbar(J_x \pm \mathrm{i}J_y)
\end{aligned} \tag{4.1.13}
$$

or

$$J_z J_\pm = J_\pm(J_z \pm \hbar) \,. \tag{4.1.14}$$

Accordingly, the two operators $J_\pm$ act as ladder operators for the quantum number m,

$$
\begin{aligned}
J_z J_\pm |l, m\rangle &= J_\pm(J_z \pm \hbar)|j, m\rangle \\
&= J_\pm |j, m\rangle \hbar(m \pm 1) \,,
\end{aligned} \tag{4.1.15}
$$

telling us that $J_\pm |j, m\rangle$ is an eigenket of J_z with eigenvalue $\hbar(m \pm 1)$. Since J_+ and J_- commute with $\vec{J}^2$,

$$\left[\vec{J}^2, J_\pm\right] = 0 \,, \tag{4.1.16}$$

we also have

$$\vec{J}^2 J_\pm |j, m\rangle = J_\pm \vec{J}^2 |j, m\rangle = J_\pm |j, m\rangle \hbar^2 j(j + 1) \,, \tag{4.1.17}$$

so that $J_\pm |j, m\rangle$ is an eigenket of $\vec{J}^2$ with eigenvalue $\hbar^2 j(j + 1)$. Taken together, these statements imply

$$
\begin{aligned}
J_+ |j, m\rangle &\propto |j, m + 1\rangle \,, \\
J_- |j, m\rangle &\propto |j, m - 1\rangle \,,
\end{aligned} \tag{4.1.18}
$$

where the factor of proportionality is to be found. We get it, of course, from the normalization of all kets $|j, m\rangle$ to unit length. So, what is the length of $J_\pm |j, m\rangle$? Let us see,

$$\langle j, m|J_\pm^\dagger J_\pm |j, m\rangle = \langle j, m|J_\mp J_\pm |j, m\rangle \tag{4.1.19}$$

with

$$J_{\mp}J_{\pm} = (J_x \mp \mathrm{i}J_y)(J_x \pm \mathrm{i}J_y)$$
$$= J_x^2 + J_y^2 \pm \mathrm{i}(J_x J_y - J_y J_x)$$
$$= J_x^2 + J_y^2 \pm \mathrm{i}(\mathrm{i}\hbar J_z)$$
$$= \vec{J}^2 - J_z^2 \mp \hbar J_z , \qquad (4.1.20)$$

so that

$$\langle j, m | J_{\pm}^{\dagger} J_{\pm} | j, m \rangle = \hbar^2 [j(j+1) - m^2 \mp m]$$
$$= \hbar^2 (j \mp m)(j \pm m + 1) , \qquad (4.1.21)$$

where the right-hand side cannot be negative. It follows that there is a last rung on the m ladder both for climbing up by applying J_+ and for climbing down by applying J_-. In the up-climb we stop at $m = j$, in the down-climb at $m = -j$, and since the steps are by changing m to $m+1$ or $m-1$, respectively, the difference $2j$ between $m = +j$ and $m = -j$ must be an integer. As a consequence, we have these possible eigenvalues for $\vec{J}^2$ and J_z:

$\vec{J}^2$ has eigenvalues $\hbar^2 j(j+1)$ with $j = 0, \frac{1}{2}, 1, \frac{3}{2}, 2, \dots$;

J_z has eigenvalues $\hbar m$ with $m = j, j-1, j-2, \dots, -j+1, -j$. (4.1.22)

These are $2j + 1$ different m values for the given j value.

With the usual convention that the normalization factors in (4.1.18) are positive, we further establish that

$$J_{\pm}|j, m\rangle = |j, m \pm 1\rangle \hbar \sqrt{(j \mp m)(j \pm m + 1)} . \qquad (4.1.23)$$

4-1 Use this to find the 3×3 matrix representations of J_x and J_y, referring to the $|j, m\rangle$ kets, for $j = 1$. Then find the eigenkets of J_x and J_y as linear superpositions of the $|j = 1, m\rangle$ kets.

4-2   Consider two independent pairs of harmonic-oscillator ladder operators, $A_1^\dagger, A_1$ and $A_2^\dagger, A_2$ with the familiar commutation relations

$$[A_j, A_k] = 0, \quad [A_j, A_k^\dagger] = \delta_{jk}, \quad [A_j^\dagger, A_k^\dagger] = 0 \quad \text{for} \quad j, k = 1, 2.$$

Show that

$$J_x = \frac{\hbar}{2}\left(A_1^\dagger A_2 + A_2^\dagger A_1\right),$$

$$J_y = \frac{\hbar}{2i}\left(A_1^\dagger A_2 - A_2^\dagger A_1\right),$$

$$J_z = \frac{\hbar}{2}\left(A_1^\dagger A_1 - A_2^\dagger A_2\right)$$

obey the commutation relations of the cartesian components of the angular momentum vector operator $\vec{J} \cong (J_x, J_y, J_z)$. Do you recognize the Pauli matrices of (2.9.9) in *Basic Matters* in these expressions for J_x, J_y, and J_z?

4-3   Now define operator J by $\vec{J}^2 = J(J+1)$ and express it in terms of the ladder operators $A_j^\dagger, A_j$. How are the joint eigenkets $|n_1, n_2\rangle$ of $A_1^\dagger A_1$ and $A_2^\dagger A_2$ related to the angular momentum kets $|j, m\rangle$?

4.2   Addition of two angular momenta

In (4.1.8), we add two angular momentum vector operators to get a third one. This raises the question of adding any two, as in

$$\vec{J} = \vec{J}_1 + \vec{J}_2. \tag{4.2.1}$$

For $\vec{J}_1$ we have the common eigenstates of $\vec{J}_1^{\,2}$ and J_{1z} with eigenvalues $\hbar^2 j_1(j_1 + 1)$ and $\hbar m_1$, respectively, and likewise for $\vec{J}_2$ with j_2 and m_2. There are altogether $(2j_1 + 1)(2j_2 + 1)$ kets of the type

$$|(j_1, m_1)(j_2, m_2)\rangle \tag{4.2.2}$$

the common eigenkets of $\vec{J}_1^{\,2}$ and J_{1z} as well as $\vec{J}_2^{\,2}$ and J_{2z}.

Since

$$J_z = J_{1z} + J_{2z} \tag{4.2.3}$$

they are also eigenkets of J_z with eigenvalue $\hbar(m_1 + m_2)$,

$$J_z|(j_1, m_1)(j_2, m_2)\rangle = |(j_1, m_1)(j_2, m_2)\rangle\hbar(m_1 + m_2) \tag{4.2.4}$$

so that the possible values of $m = m_1 + m_2$ range from $m = -(j_1 + j_2)$ to $m = j_1 + j_2$. The largest value $m = j_1 + j_2$ is only realized for $m_1 = j_1$ and $m_2 = j_2$, but there are two possibilities for $m = j_1 + j_2 - 1$, namely $(m_1, m_2) = (j_1, j_2 - 1)$ and $(j_1 - 1, j_2)$; there are three possibilities for $m = j_1 + j_2 - 2$, and so forth. These matters are summarized in the following table:

m	(m_1, m_2) pairs	dimension of subspace
$j_1 + j_2$	(j_1, j_2)	1
$j_1 + j_2 - 1$	$(j_1, j_2 - 1), (j_1 - 1, j_2)$	2
$j_1 + j_2 - 2$	$(j_1, j_2 - 2), (j_1 - 1, j_2 - 1), (j_1 - 2, j_2)$	3
$\vdots$	$\vdots$	$\vdots$
$-j_1 - j_2 + 1$	$(-j_1 + 1, -j_2), (-j_1, -j_2 + 1)$	2
$-j_1 - j_2$	$(-j_1, -j_2)$	1

The largest possible value for j — eigenvalue $\hbar^2 j(j+1)$ of $\vec{J}^2$ — is, therefore, $j = j_1 + j_2$. We write

$$\left| j_1, j_2; j, m \right\rangle \tag{4.2.5}$$

for the common eigenkets of $\vec{J}^2$ and J_z, indicating that these $\left| j, m \right\rangle$ kets come about by adding $\vec{J}_1$ and $\vec{J}_2$ with j_1 and j_2 as the quantum numbers for $\vec{J}_1^{\,2}$ and $\vec{J}_2^{\,2}$. Clearly,

$$\left| j_1, j_2; j_1 + j_2, j_1 + j_2 \right\rangle = \left| (j_1, j_1)(j_2, j_2) \right\rangle \tag{4.2.6}$$

for the state with $j = j_1 + j_2$ and $m = j$. The other states to $j = j_1 + j_2$ are then obtained by successive applications of

$$J_- = J_{1-} + J_{2-} = (J_{1x} + J_{2x}) - \mathrm{i}(J_{1y} + J_{2y}), \tag{4.2.7}$$

giving

$$\left| j_1, j_2; j = j_1 + j_2, m \right\rangle \propto (J_-)^{j_1 + j_2 - m} \left| j_1, j_2; j = j_1 + j_2, m = j \right\rangle, \tag{4.2.8}$$

where the proportionality factor is the respective product of the square-root factors in (4.1.23) (or rather their reciprocals).

For each m value this gives a particular linear combination of the kets $|(j_1, m_1)(j_2, m_2)\rangle$ with $m_1 + m_2 = m$. For example, we have

$$|j_1, j_2; j = j_1 + j_2, m = j - 1\rangle$$

$$= \frac{(J_{1-} + J_{2-})|(j_1, m_1 = j_1)(j_2, m_2 = j_2)\rangle}{\sqrt{2j}\,\hbar}$$

$$= |(j_1, m_1 = j_1 - 1)(j_2, m_2 = j_2)\rangle \frac{\sqrt{2j_1}}{\sqrt{2j}}$$

$$+ |(j_1, m_1 = j_1)(j_2, m_2 = j_2 - 1)\rangle \frac{\sqrt{2j_2}}{\sqrt{2j}}. \quad (4.2.9)$$

The orthogonal linear combination of the m_1, m_2 kets, with coefficients $\sqrt{j_2/j}$ and $-\sqrt{j_1/j}$ is what is left over in the two-dimensional subspace to $m = j_1 + j_2 - 1$. So, this second state must be the joint eigenstate of $\vec{J}^2$ and J_z with eigenvalues corresponding to the quantum numbers $j = j_1 + j_2 - 1$ and $m = j$,

$$|j_1, j_2; j = m = j_1 + j_2 - 1\rangle$$

$$= |(j_1, m_1 = j_1)(j_2, m_2 = j_2 - 1)\rangle \sqrt{\frac{j_1}{j_1 + j_2}}$$

$$- |(j_1, m_1 = j_1 - 1)(j_2, m_2 = j_2)\rangle \sqrt{\frac{j_2}{j_1 + j_2}}, \quad (4.2.10)$$

and we apply $J_- = J_{1-} + J_{2-}$ repeatedly to get the kets for all other m values. Thereafter, we have the all kets $|j, m\rangle$ for $j = j_1 + j_2$ and $j = j_1 + j_2 - 1$ and all m values. At this stage, all kets with $m = \pm(j_1 + j_2)$ and $m = \pm(j_1 + j_2 - 1)$ have been used up, and there is one linear combination left for $m_1 + m_2 = j_1 + j_2 - 2$, and one for $m_1 + m_2 = -(j_1 + j_2 - 2)$. These must then be the states for $j = j_1 + j_2 - 2$ and $m = \pm j$, and beginning with the $m = +j$ state, say, we get all others by applying J_- repeatedly. And so forth: we get successively the kets $|jm\rangle$ for $j = j_1 + j_2$, then $j = j_1 + j_2 - 1$, then $j = j_1 + j_2 - 2, \ldots$ until all states are used up, which happens when we reach $j = |j_1 - j_2|$, the smallest possible j value.

That this is indeed the smallest value is easily seen by a count of all states,

$$\sum_{j=|j_1 - j_2|}^{j_1 + j_2} (2j + 1) = (2j_1 + 1)(2j_2 + 1), \quad (4.2.11)$$

which, as we know, is the total number of states available. For the evaluation of this sum it helps to note that it is of the telescoping kind because $2j + 1 = (j + 1)^2 - j^2$.

In summary, the possible values for j are

$$j = j_1 + j_2, \ j_1 + j_2 - 1, \ j_1 + j_2 - 2, \ \ldots, \ |j_1 - j_2|, \qquad (4.2.12)$$

and for each j value there are, of course, $2j + 1$ values for m, namely $m = j, j - 1, \ldots, -j$. The construction discussed above, where we begin with $j = m = j_1 + j_2$, apply J_- repeatedly, then repeat the whole procedure starting with $j = m = j_1 + j_2 - 1$, until all possibilities are exhausted, tells us the coefficients in

$$|j_1, j_2; j, m\rangle = \sum_{m_1, m_2} |(j_1, m_1)(j_2, m_2)\rangle \underbrace{\langle (j_1, m_1)(j_2, m_2)|j_1, j_2; j, m\rangle}_{\text{C-G coefficient}}.$$

$$(4.2.13)$$

These so-called *Clebsch–Gordan coefficients* (Rudolf F. A. Clebsch and Paul A. Gordan, that is) can thus be calculated in a rather simple, systematic, but somewhat tedious way. They have been tabulated, and are thus readily available, and their properties have been studied diligently. We note that

$$\langle (j_1, m_1)(j_2, m_2)|j_1, j_2; j, m\rangle = 0$$

$$\text{unless} \quad m = m_1 + m_2 \quad \text{and} \quad j_1 + j_2 \geq j \geq |j_1 - j_2|, \qquad (4.2.14)$$

a basic property of the Clebsch–Gordan coefficients that follows immediately from the arguments given above. Further, most people (but not all) use the convention that the coefficient is positive for $m = j$ and $m_1 = j_1$, which is the choice we made in (4.2.10), and also in (4.2.6). With this convention, the values of the Clebsch–Gordan coefficients are unique.

4.2.1   Two spin-$\frac{1}{2}$ systems

Perhaps the simplest situation is the important case of adding two $j = \frac{1}{2}$ angular momenta. If we denote the $m = \pm\frac{1}{2}$ states by $|\uparrow\rangle$ and $|\downarrow\rangle$ for an individual $j = \frac{1}{2}$ system (see Sections 2.18 and 3.9 in *Basic Matters*, for example), then we have

$$m = m_1 + m_2 = 1 : \quad |\uparrow\uparrow\rangle \text{ only, for which } m_1 = m_2 = \tfrac{1}{2} ;$$

$$m = 0 : \quad |\uparrow\downarrow\rangle \text{ and } |\downarrow\uparrow\rangle, \text{ for which}$$
$$(m_1, m_2) = \left(\tfrac{1}{2}, -\tfrac{1}{2}\right) \text{ and } \left(-\tfrac{1}{2}, \tfrac{1}{2}\right) ;$$

$$m = -1 : \quad |\downarrow\downarrow\rangle \text{ only, for which } m_1 = m_2 = -\tfrac{1}{2}. \quad (4.2.15)$$

The procedure explained above then gives the three *triplet* states

$$|j = 1, m = 1\rangle = |\uparrow\uparrow\rangle,$$
$$|j = 1, m = 0\rangle = \left(|\uparrow\downarrow\rangle + |\downarrow\uparrow\rangle\right)/\sqrt{2},$$
$$|j = 1, m = -1\rangle = |\downarrow\downarrow\rangle \quad\quad\quad\quad (4.2.16)$$

for $j = 1$, as well as the *singlet* state

$$|j = 0, m = 0\rangle = \left(|\uparrow\downarrow\rangle - |\downarrow\uparrow\rangle\right)/\sqrt{2} \quad\quad (4.2.17)$$

for $j = 0$. The amplitude factors of $+1$ or $\pm\dfrac{1}{\sqrt{2}}$ on the right-hand sides of (4.2.16) and (4.2.17) are the nonvanishing Clebsch–Gordan coefficients for $j_1 = j_2 = \tfrac{1}{2}$ and $j = 1$ as well as $j = 0$.

The triplet states are symmetric under the exchange of the two individual spin-$\tfrac{1}{2}$ particles, whereas the singlet state is antisymmetric. In passing, we also note that the singlet state appears in (2.18.9) of *Basic Matters*.

4-4 Determine the Clebsch–Gordan coefficients for $j_1 = j_2 = 1$.

4.2.2 Total angular momentum of an electron

Another situation that is quite important is that of a single electron in a spherically symmetric potential, such as the electron in a hydrogen-like atom, or the valence electron in an alkaline atom. Such an electron has orbital angular momentum $\vec{L} = \vec{R} \times \vec{P}$ and spin-$\tfrac{1}{2}$, $\vec{S} = \dfrac{\hbar}{2}\vec{\sigma}$, so that $\vec{J} = \vec{L} + \vec{S}$ takes on values corresponding to the addition of two angular momenta with $j_1 = l$ and $j_2 = \tfrac{1}{2}$. Thus we have the following cases

quantum numbers		spectroscopic	
l	j	denotation	multiplicity
0	1/2	$s_{1/2}$	2 (doublet)
1	1/2	$p_{1/2}$	2 (doublet)
	3/2	$p_{3/2}$	4 (quartet)
2	3/2	$d_{3/2}$	4 (quartet)
	5/2	$d_{5/2}$	6 (sextet)
3	5/2	$f_{5/2}$	6 (sextet)
	7/2	$f_{7/2}$	8 (octet)

and so forth. The letters s, p, d, f stand for $l = 0, 1, 2, 3$ and the subscripts denote the value of j. In spectroscopy, one would label spectroscopic lines by identifying them as "the $d_{5/2}$ to $p_{3/2}$ transition", for example.

In hydrogen, or hydrogen-like atoms, a further label is the principal quantum number n that identifies the Bohr shell, with $n = 1, 2, 3, \ldots$, as discussed in Section 5.1 of *Simple Systems*. The possible values of l are then $l = 0, 1, 2, \ldots, n-1$ in the nth Bohr shell. In the 1st Bohr shell, $n = 1$, there is only the $s_{1/2}$ doublet: $1s_{1/2}$; in the 2nd Bohr shell, we have $2s_{1/2}, 2p_{1/2}$ and $2p_{3/2}$; in the 3rd Bohr shell, there are $3s_{1/2}, 3p_{1/2}, 3p_{3/2}, 3d_{3/2}$ and $3d_{5/2}$, and so forth. An even finer classification would take into account that the nucleus may also have spin and thus contribute to the total angular momentum of the atom.

Chapter 5

External Magnetic Field

5.1 Electric charge in a magnetic field

A charge q moving in a magnetic field $\vec{B}(\vec{R})$ experiences a velocity-dependent *Lorentz force*, named after Hendrik A. Lorentz,

$$M\frac{\mathrm{d}^2}{\mathrm{d}t^2}\vec{R}(t) = \frac{q}{c}\frac{\mathrm{d}\vec{R}}{\mathrm{d}t} \times \vec{B}(\vec{R}),\qquad(5.1.1)$$

where c is the speed of light. A Hamilton operator for this physical situation cannot be of the typical structure

$$H = \frac{1}{2M}\vec{P}^2 + V(\vec{R}),\qquad(5.1.2)$$

because this does not give rise to velocity-dependent forces. Rather, we need a structurally different Hamilton operator, actually of the form

$$H = \frac{1}{2M}\left(\vec{P} - \frac{q}{c}\vec{A}(\vec{R})\right)^2,\qquad(5.1.3)$$

where $\vec{A}(\vec{R})$ is a vector potential for the magnetic field,

$$\vec{B}(\vec{R}) = \vec{\nabla} \times \vec{A}(\vec{r}).\qquad(5.1.4)$$

For simplicity, we consider only the case of a time-independent vector potential and thus time-independent magnetic field but, except for slight additional complications in the following formulas, matters remain largely the same even if a parametric time dependence is present in $\vec{B}$ and $\vec{A}$.

For this Hamilton operator we have the velocity operator

$$\vec{V} = \frac{\mathrm{d}}{\mathrm{d}t}\vec{R} = \frac{\partial H}{\partial \vec{P}} = \frac{1}{M}\left(\vec{P} - \frac{q}{c}\vec{A}(\vec{R})\right).\qquad(5.1.5)$$

The first thing to note is that now the momentum $\vec{P}$ is not identical with $M\vec{V}$, the so-called *kinetic momentum*. One sometimes refers to $\vec{P}$ as the *canonical momentum* when one wishes to emphasize the difference between $\vec{P}$ and $M\vec{V}$.

Secondly, we note that

$$H = \frac{M}{2}\vec{V}^2 \qquad (5.1.6)$$

with this expression for the velocity. Since there is no parametric time dependence in $\vec{A}(\vec{R})$, there is no parametric time dependence in H, and it follows that $\dfrac{dH}{dt} = 0$. Therefore, the change of the velocity $\vec{V}$ in time can only be a change in the direction, not a change of the speed $|\vec{V}|$. As expected, we see here that a charge moves in a magnetic field with constant speed but changes direction in time.

Before we can really justify these conclusions, we must demonstrate that the correct equations of motion emerge from the Hamilton operator (5.1.3). We need to consider

$$\frac{d}{dt}\left(M\vec{V}\right) = \frac{1}{i\hbar}\left[M\vec{V}, H\right] = \frac{M^2}{2i\hbar}\left[\vec{V}, \vec{V}^2\right], \qquad (5.1.7)$$

which raises the issue of the commutation relations among components of the velocity operator $\vec{V}$.

With numerical vectors $\vec{a}$ and $\vec{b}$ we have, as an application of (1.2.92), or rather its three-dimensional version in (4.1.7) of *Simple Systems*,

$$
\begin{aligned}
\frac{1}{i\hbar}\left[\vec{a}\cdot\vec{V}, \vec{b}\cdot\vec{V}\right] &= \frac{1}{i\hbar}\frac{1}{M^2}\left[\vec{a}\cdot\vec{P} - \frac{q}{c}\vec{a}\cdot\vec{A}, \vec{b}\cdot\vec{P} - \frac{q}{c}\vec{b}\cdot\vec{A}\right] \\
&= \frac{q}{M^2 c}\left(\vec{a}\cdot\vec{\nabla}\vec{b}\cdot\vec{A} - \vec{b}\cdot\vec{\nabla}\vec{a}\cdot\vec{A}\right) \\
&= \frac{q}{M^2 c}\left(\vec{a}\times\vec{b}\right)\cdot\left(\vec{\nabla}\times\vec{A}\right) \\
&= \frac{q}{M^2 c}\left(\vec{a}\times\vec{b}\right)\cdot\vec{B}, \qquad (5.1.8)
\end{aligned}
$$

stating that the commutator between two components of this velocity vector operator is proportional to the component of the magnetic field in the third direction. As a consequence of the commutator we have

$$
\begin{aligned}
\frac{1}{i\hbar}\left[\vec{a}\cdot\vec{V}, \vec{V}^2\right] &= \frac{q}{M^2 c}\left(\vec{a}\cdot\left(\vec{b}\times\vec{B}\right)\Big|_{\vec{b}=\vec{V}} - \vec{a}\cdot\left(\vec{B}\times\vec{b}\right)\Big|_{\vec{b}=\vec{V}}\right) \\
&= \frac{q}{M^2 c}\vec{a}\cdot\left(\vec{V}\times\vec{B} - \vec{B}\times\vec{V}\right), \qquad (5.1.9)
\end{aligned}
$$

and we get

$$\frac{d}{dt}\left(M\vec{V}\right) = \frac{q}{2c}\left(\vec{V}\times\vec{B} - \vec{B}\times\vec{V}\right),\tag{5.1.10}$$

which is the properly symmetrized version of the Lorentz force in (5.1.1).

5-1 Work out the difference between $\frac{1}{2}\left(\vec{V}\times\vec{B} - \vec{B}\times\vec{V}\right)$ and $\vec{V}\times\vec{B}$.

5-2 Modify the respective expressions of Section 3.1 to derive the probability current density for the Hamilton operator (5.1.3), which is *not* of the form (3.1.8).

Having thus demonstrated that

$$H = \frac{1}{2M}\left(\vec{P} - \frac{q}{c}\vec{A}(\vec{R})\right)^2 \quad\text{with}\quad \vec{\nabla}\times\vec{A}(\vec{r}) = \vec{B}(\vec{r})\tag{5.1.11}$$

is the correct Hamilton operator, indeed, let us now apply it to the situation of a homogeneous magnetic field in the z direction,

$$\vec{B} = B\vec{e}_z \,\hat{=}\, (0,0,B)\,.\tag{5.1.12}$$

Then the three cartesian components of $\vec{V} \,\hat{=}\, (V_1, V_2, V_3)$ obey the commutation relations

$$[V_1, V_2] = i\hbar\frac{qB}{M^2c}\,,\quad [V_1, V_3] = 0\,,\quad [V_2, V_3] = 0\,,\tag{5.1.13}$$

and the Hamilton operator

$$H = \underbrace{\frac{M}{2}\left(V_1^2 + V_2^2\right)}_{=H_\perp} + \underbrace{\frac{M}{2}V_3^2}_{=H_\parallel} = H_\perp + H_\parallel\tag{5.1.14}$$

splits naturally into two dynamically independent parts: the force-free motion along the z direction, which is of no further interest here, and the accelerated but speed-conserving motion in the xy plane governed by $H_\perp$.

We recognize that the pair V_1, V_2 is much like a position-momentum pair, having a commutator that is a multiple of the identity. It will be expedient to introduce nonhermitian operators ($qB > 0$ assumed)

$$A = \sqrt{\frac{M^2c}{2\hbar qB}}(V_1 + iV_2)\,,$$

$$A^\dagger = \sqrt{\frac{M^2c}{2\hbar qB}}(V_1 - iV_2)\,,\tag{5.1.15}$$

which obey

$$[A, A^\dagger] = 1, \tag{5.1.16}$$

the commutation relation for the ladder operators of a harmonic oscillator. Then

$$H_\perp = \frac{\hbar q B}{Mc}\left(A^\dagger A + \frac{1}{2}\right) = \hbar\omega_{\text{cycl}}\left(A^\dagger A + \frac{1}{2}\right), \tag{5.1.17}$$

that is: $H_\perp$ is the Hamilton operator of a harmonic oscillator with the *cyclotron frequency* (remember that $qB > 0$ is assumed)

$$\omega_{\text{cycl}} = \frac{qB}{Mc}. \tag{5.1.18}$$

Therefore, the eigenvalues of $H_\perp$ are

$$\frac{1}{2}\hbar\omega_{\text{cycl}}, \; \frac{3}{2}\hbar\omega_{\text{cycl}}, \; \frac{5}{2}\hbar\omega_{\text{cycl}}, \; \dots, \tag{5.1.19}$$

but — contrary to the one-dimensional harmonic oscillator — these eigenvalues are highly degenerate. For, what is determined by stating the energy are the expectation values of $A, A^\dagger, A^\dagger A, \dots$, all referring solely to the velocity of the charged particle, but not at all to its position.

To illuminate this matter, let us consider two particular choices for the vector potential $\vec{A}$, namely

    (1) the symmetric choice $\vec{A} = \dfrac{1}{2}\vec{B} \times \vec{R} \cong \left(-\frac{1}{2}BX_2, \frac{1}{2}BX_1, 0\right)$,

    (2) the asymmetric choice $\vec{A} = BX_1\vec{e}_y \cong (0, BX_1, 0)$. $\qquad$ (5.1.20)

One verifies easily that

$$\vec{\nabla} \times \vec{A} = B\vec{e}_z \cong (0, 0, B) \tag{5.1.21}$$

for both choices.

For the symmetric choice in (5.1.20), we have

$$\begin{aligned}
H_\perp &= \frac{M}{2}\left(V_1^2 + V_2^2\right) \\
&= \frac{M}{2}\left(\frac{1}{M}P_1 + \frac{1}{2}\omega_{\text{cycl}}X_2\right)^2 + \frac{M}{2}\left(\frac{1}{M}P_2 - \frac{1}{2}\omega_{\text{cycl}}X_1\right)^2 \\
&= \frac{1}{2M}\left(P_1^2 + P_2^2\right) + \frac{1}{2}M\left(\frac{\omega_{\text{cycl}}}{2}\right)^2\left(X_1^2 + X_2^2\right) \\
&\quad - \frac{\omega_{\text{cycl}}}{2}\left(X_1 P_2 - X_2 P_1\right),
\end{aligned} \tag{5.1.22}$$

which is the Hamilton operator of a two-dimensional harmonic oscillator with natural frequency $\frac{1}{2}\omega_{\text{cycl}}$ and an extra term proportional to

$$L_z = X_1 P_2 - X_2 P_1 \,. \tag{5.1.23}$$

Recalling the lessons of Section 3.5 in *Simple Systems*, we note that it is fitting to switch to the $A_\pm$ ladder operators,

$$A_\pm^\dagger = \frac{1}{\sqrt{2M\hbar\omega_{\text{cycl}}}} \left(\frac{1}{2} M\omega_{\text{cycl}}(X_1 \pm iX_2) - i(P_1 \pm iP_2) \right),$$

$$A_\pm = \frac{1}{\sqrt{2M\hbar\omega_{\text{cycl}}}} \left(\frac{1}{2} M\omega_{\text{cycl}}(X_1 \mp iX_2) + i(P_1 \mp iP_2) \right), \tag{5.1.24}$$

for which

$$H_\perp = \frac{1}{2}\hbar\omega_{\text{cycl}} \left(A_+^\dagger A_+ + A_-^\dagger A_- + 1 \right) - \frac{1}{2}\hbar\omega_{\text{cycl}} \underbrace{\left(A_+^\dagger A_+ - A_-^\dagger A_- \right)}_{= L_z/\hbar} \tag{5.1.25}$$

so that

$$H_\perp = \hbar\omega_{\text{cycl}} \left(A_-^\dagger A_- + \frac{1}{2} \right). \tag{5.1.26}$$

That is: Although there are two harmonic oscillators (ladder operators $A_+^\dagger, A_+$ for one and the pair $A_-^\dagger, A_-$ for the other) only one of them is dynamically relevant; the "+" oscillator does not appear in the Hamilton operator. The energy eigenvalues of the kets $|n_+, n_-\rangle$,

$$H_\perp |n_+, n_-\rangle = |n_+, n_-\rangle \hbar\omega_{\text{cycl}} \left(n_- + \frac{1}{2}\right) \tag{5.1.27}$$

do not depend on the quantum number n_+ and are, therefore, infinitely degenerate.

What is the physical origin of this degeneracy? To answer this question, we return to the equation of motion (5.1.10) and write it for the relevant perpendicular part $\vec{V}_\perp \cong (V_1, V_2, 0)$ for the present case of a homogeneous magnetic field in the third direction, $\vec{B} \cong (0, 0, B)$, namely

$$\frac{d}{dt}\left(M\vec{V}_\perp\right) = \frac{q}{c}\vec{V}_\perp \times \vec{B} \,. \tag{5.1.28}$$

With

$$\vec{V}_\perp = \frac{d}{dt}\vec{R}_\perp \,, \quad \vec{R}_\perp \cong (X_1, X_2, 0) \tag{5.1.29}$$

this reads

$$\frac{d}{dt}\left(M\vec{V}_\perp + \frac{q}{c}\vec{B} \times \vec{R}_\perp\right) = 0.\tag{5.1.30}$$

Accordingly, what is differentiated here is a constant of motion, for which it will be expedient to write

$$M\vec{V}_\perp(t) + \frac{q}{c}\vec{B} \times \vec{R}_\perp(t) = \frac{q}{c}\vec{B} \times \vec{R}_0\tag{5.1.31}$$

where $\vec{R}_0 \mathrel{\widehat{=}} (X_0, Y_0, 0)$ is a time-independent vector operator in the xy plane. We thus have

$$\vec{V}_\perp(t) = -\frac{q}{Mc}\vec{B} \times \left(\vec{R}_\perp(t) - \vec{R}_0\right),\tag{5.1.32}$$

which — as it should — describes the circular motion around position $\vec{R}_0$ with the angular velocity $qB/(Mc) = \omega_{\text{cycl}}$.

We make explicit the two components in the perpendicular xy plane,

$$V_1 = \omega_{\text{cycl}}(X_2 - Y_0), \qquad V_2 = -\omega_{\text{cycl}}(X_1 - X_0),\tag{5.1.33}$$

and recall that

$$V_1 = \frac{1}{M}P_1 + \frac{1}{2}\omega_{\text{cycl}}X_2, \qquad V_2 = \frac{1}{M}P_2 - \frac{1}{2}\omega_{\text{cycl}}X_1\tag{5.1.34}$$

for the symmetric choice. Accordingly, we have

$$X_0 = X_1 + \frac{1}{\omega_{\text{cycl}}}V_2 = \frac{1}{2}X_1 + \frac{1}{M\omega_{\text{cycl}}}P_2,$$

$$Y_0 = X_2 - \frac{1}{\omega_{\text{cycl}}}V_1 = \frac{1}{2}X_2 - \frac{1}{M\omega_{\text{cycl}}}P_1\tag{5.1.35}$$

for the coordinates of the center in the xy plane around which the circular motion happens. But, in view of

$$[X_0, Y_0] = -\frac{i\hbar}{M\omega_{\text{cycl}}}\tag{5.1.36}$$

the two coordinates cannot be specified simultaneously, their spreads obey the Heisenberg-type uncertainty relation

$$\delta X_0\,\delta Y_0 \geq \frac{\hbar}{2M\omega_{\text{cycl}}}.\tag{5.1.37}$$

Now consider the square of $\vec{R}_0$, the squared distance of the center of the circular motion from the origin of the coordinate system,

$$
\begin{aligned}
\vec{R}_0^2 = X_0^2 + Y_0^2 &= \frac{1}{4}(X_1^2 + X_2^2) + \left(\frac{1}{M\omega_{\text{cycl}}}\right)^2 (P_1^2 + P_2^2) \\
&\quad + \frac{1}{M\omega_{\text{cycl}}}(X_1 P_2 - X_2 P_1) \\
&= \frac{2}{M\omega_{\text{cycl}}^2}\left[\frac{1}{2M}(P_1^2 + P_2^2) + \frac{M}{2}\left(\frac{\omega_{\text{cycl}}}{2}\right)^2 (X_1^2 + X_2^2)\right] \\
&\quad + \frac{1}{M\omega_{\text{cycl}}}L_z .
\end{aligned}
\tag{5.1.38}
$$

After introducing the $A_\pm$ ladder operators of (5.1.24) this reads

$$
\begin{aligned}
X_0^2 + Y_0^2 &= \frac{2}{M\omega_{\text{cycl}}^2}\frac{\hbar\omega_{\text{cycl}}}{2}\left(A_+^\dagger A_+ + A_-^\dagger A_- + 1\right) \\
&\quad + \frac{\hbar}{M\omega_{\text{cycl}}}\left(A_+^\dagger A_+ - A_-^\dagger A_-\right) \\
&= \frac{2\hbar}{M\omega_{\text{cycl}}}\left(A_+^\dagger A_+ + \frac{1}{2}\right).
\end{aligned}
\tag{5.1.39}
$$

It follows that the ket $|n_+, n_-\rangle$ is not only an eigenket of $H_\perp$ with energy eigenvalue $\hbar\omega_{\text{cycl}}(n_- + \frac{1}{2})$, see (5.1.27), but also an eigenket of $X_0^2 + Y_0^2$ with eigenvalue $2\hbar/(M\omega_{\text{cycl}})(n_+ + \frac{1}{2})$. This is to say: The distance of the center of the circular motion from the coordinate origin in the xy plane is specified precisely to be

$$
r_0 = \sqrt{\frac{2\hbar}{M\omega_{\text{cycl}}}\left(n_+ + \frac{1}{2}\right)},
\tag{5.1.40}
$$

but we do not know where, on this circle of radius r_0, the center is located.

This center can be anywhere. With the symmetric choice for the vector potential, option (1) in (5.1.20), we specify $X_0^2 + Y_0^2$, but the individual values of X_0 and Y_0 are subject to the uncertainty relation in (5.1.37).

5-3 The radius of the circle of the circular motion in the xy plane is the square root of

$$\left(\vec{R}_\perp(t) - \vec{R}_0\right)^2.$$

Does it have a definite value if ket $|n_+, n_-\rangle$ describes the state of the system?

5-4 For the asymmetric choice (2) in (5.1.20), find V_1, V_2 as well as $H_\perp$ and X_0, Y_0 in terms of X_1, X_2, P_1, and P_2. Verify that the commutation relations of (5.1.13) and (5.1.36) are obeyed.

5-5 Show that

$$[V_1, X_0] = [V_2, X_0] = [V_1, Y_0] = [V_2, Y_0] = 0$$

for *any* vector potential $\vec{A}(\vec{R})$ to $\vec{B} = B\,\vec{e}_z$, and verify also that the commutators $[V_1, V_2]$ and $[X_0, Y_0]$ have the values of (5.1.13) and (5.1.36).

5-6 Now construct, quite generally, joint eigenkets $|n_1, n_2\rangle$ of $H_\perp$ and $X_0^2 + Y_0^2$ such that

$$H_\perp |n_1, n_2\rangle = |n_1, n_2\rangle \hbar\omega_{\text{cycl}}\left(n_1 + \frac{1}{2}\right)$$

and

$$(X_0^2 + Y_0^2)|n_1, n_2\rangle = |n_1, n_2\rangle \frac{2\hbar}{M\omega_{\text{cycl}}}\left(n_2 + \frac{1}{2}\right).$$

5.2 Electron in a homogeneous magnetic field

In addition to carrying one negative unit of charge, $q = -e$ with $e > 0$, an electron has a magnetic dipole moment $\vec{\mu}$. Since this is a vector, it must be proportional to the only other intrinsic vector available for the electron, the spin vector $\vec{S} = \frac{\hbar}{2}\vec{\sigma}$. So we write

$$\vec{\mu} = -g\mu_{\text{B}}\vec{S}/\hbar = -\frac{1}{2}g\mu_{\text{B}}\vec{\sigma} \tag{5.2.1}$$

where the minus sign is appropriate for the negatively charged electron and μ_{B} is the Bohr magneton, named after Niels H. D. Bohr,

$$\mu_{\text{B}} = \frac{e\hbar}{2Mc} = 5.788 \times 10^{-9}\,\text{eV/G} \tag{5.2.2}$$

($M = 9.1094 \times 10^{-28}$ g is the electron mass, $e = 4.803 \times 10^{-10}$ esu is the elementary charge). The proportionality factor g is the *gyromagnetic ratio*, or "g-factor", which is very close to 2 for the electron. Indeed, we shall take $g = 2$ for the purpose of the present discussion and ignore that the correct value is about 0.1% larger. This so-called anomaly of the electron g-factor originates in subtle relativistic effects which are well beyond the scope of these lectures.

There is then an additional magnetic interaction energy to be taken into account,

$$H_{\text{magn}} = -\vec{\mu} \cdot \vec{B} = g\mu_{\text{B}} \vec{S} \cdot \vec{B}/\hbar \tag{5.2.3}$$

which supplements the Hamilton operator in (5.1.3) to give the complete Hamilton operator ($q \to -e$ now)

$$H = \frac{1}{2M}\left[\vec{P} + \frac{e}{c}\vec{A}(\vec{R})\right]^2 + g\mu_{\text{B}} \vec{S} \cdot \vec{B}. \tag{5.2.4}$$

We restrict the discussion to the situation of a homogeneous magnetic field and choose the vector potential in the symmetric form (1) of (5.1.20),

$$\vec{A} = \frac{1}{2}\vec{B} \times \vec{R}. \tag{5.2.5}$$

The kinetic energy is then

$$
\begin{aligned}
H_{\text{kin}} = \frac{M}{2}\vec{V}^2 &= \frac{1}{2M}\left(\vec{P} + \frac{e}{2c}\vec{B} \times \vec{R}\right)^2 \\
&= \frac{1}{2M}\vec{P}^2 + \frac{e}{4Mc}\left[\vec{P} \cdot \left(\vec{B} \times \vec{R}\right) + \left(\vec{B} \times \vec{R}\right) \cdot \vec{P}\right] \\
&\quad + \frac{1}{2M}\left(\frac{e}{2c}\right)^2\left(\vec{B} \times \vec{R}\right)^2.
\end{aligned}
\tag{5.2.6}
$$

The terms linear in the magnetic field $\vec{B}$ can be rewritten to exhibit the orbital angular momentum $\vec{L} = \vec{R} \times \vec{P}$,

$$
\begin{aligned}
\vec{P} \cdot \left(\vec{B} \times \vec{R}\right) + \left(\vec{B} \times \vec{R}\right) \cdot \vec{P} &= \left(-\vec{P} \times \vec{R} + \vec{R} \times \vec{P}\right) \cdot \vec{B} \\
&= 2\vec{L} \cdot \vec{B},
\end{aligned}
\tag{5.2.7}
$$

so that, with the inclusion of H_{magn} of (5.2.3),

$$H = \frac{1}{2M}\vec{P}^2 + \frac{\mu_{\text{B}}}{\hbar}\left(\vec{L} + g\vec{S}\right) \cdot \vec{B} + \frac{e^2}{8Mc^2}\left(\vec{B} \times \vec{R}\right). \tag{5.2.8}$$

Note the appearance of the sum $\vec{L} + g\vec{S}$, which tells us that the magnetic moment associated with the orbital motion of the electron, that is: the magnetic moment that stems from the electric current of the moving electron, has the size of one Bohr magneton μ_B per $\hbar$, the unit of angular momentum. By contrast, the magnetic moment of the spin angular momentum is amplified by $g \cong 2$; it is essentially twice as large. Put differently, the spin of $\frac{1}{2}\hbar$ gives rise to a magnetic interaction as strong as that for orbital angular momentum of $\hbar$.

Aiming at a perturbative 1st-order treatment of the energy shifts introduced by the magnetic field we neglect the term proportional to $\vec{B}^2$ (in atoms, it gives rise to diamagnetism), and get the 1st-order effect by applying the Hellmann–Feynman theorem of Section 6.1 in *Simple Systems*, or any equivalent statement of perturbation theory. In short, we have

$$\delta E_{\text{magn}} = \frac{\mu_B}{\hbar} \left\langle \left(\vec{L} + g\vec{S} \right) \cdot \vec{B} \right\rangle , \qquad (5.2.9)$$

wherein the expectation value is evaluated for the unperturbed states.

As derived, this applies to a single electron in a homogeneous magnetic field. But just as well there could have been an additional potential energy, such as the Coulombic interaction with an atomic nucleus, and there could be more than one electron. Then $\vec{L}$ and $\vec{S}$ stand for the total orbital and spin angular momenta, the sums of the contributions form the individual electrons,

$$\vec{L} = \sum_j \vec{L}_j , \quad \vec{S} = \sum_j \vec{S}_j \qquad (5.2.10)$$

with $\vec{L}_j, \vec{S}_j$ referring to the jth electron.

We take for granted that if there is another potential energy, it is spherically symmetric, as is the case for the Coulomb potential in an atom. Then the unperturbed states can either be chosen as

$$\text{joint eigenstates } \left| (l, m_l)(s, m_s) \right\rangle \text{ of } \vec{L}^2, L_z, \vec{S}^2, \text{ and } S_z \qquad (5.2.11)$$

(for $\vec{B} = B\vec{e}_z$, that is $\vec{B}$ is aligned with the z axis) or we can choose

$$\text{joint eigenstates } \left| l, s; j, m_j \right\rangle \text{ of } \vec{L}^2, \vec{S}^2, \vec{J}^2, \text{ and } J_z \qquad (5.2.12)$$

where $\vec{J} = \vec{L} + \vec{S}$ is the total angular momentum.

For the L_z, S_z set of (5.2.11), we have

$$\delta E_{\text{magn}} = (m_l + g m_s)\mu_B B , \qquad (5.2.13)$$

where $m_l = 0, \pm 1, \pm 2, \ldots$ is always integer, whereas m_s is integer for an even number of electrons, half-integer for an odd number of electrons. But for $g = 2$, the product gm_s is always integer as well. Therefore, these magnetic energy shifts are degenerate. For example, if there are three electrons, so that $m_s = \pm\frac{1}{2}, \pm\frac{3}{2}$ are the possible values for m_s, we get the same value for δE_{magn} for the quantum numbers

$$(m_1, m_s) = \left(0, \tfrac{1}{2}\right), \left(2, -\tfrac{1}{2}\right), \left(-2, \tfrac{3}{2}\right), \left(4, -\tfrac{3}{2}\right), \qquad (5.2.14)$$

that is $\delta E_{\text{magn}} = \mu_{\text{B}} B$ is fourfold degenerate here.

Matters are remarkably more involved when we choose the $\vec{J}^2, J_z$ states of (5.2.12) as the unperturbed set. We need a special case of the Wigner–Eckart theorem (named after Eugene P. Wigner and Carl Eckart) to handle this. So, as a preparation, consider the three operators defined by

$$Z_0 \equiv S_z, \quad Z_{\pm 1} = \frac{1}{\sqrt{2}}(\mp S_x - iS_y). \qquad (5.2.15)$$

One verifies easily that

$$[J_\pm, Z_\alpha] = \hbar\sqrt{(1 \mp \alpha)(2 \pm \alpha)}\, Z_{\alpha\pm 1} \qquad (5.2.16)$$

where $J_\pm = J_x \pm iJ_y$ are the familiar angular momentum ladder operators, which have an orbital and a spin part,

$$J_\pm = (L_x \pm iL_y) + (S_x \pm iS_y). \qquad (5.2.17)$$

5-7   Verify (5.2.16) indeed.

We use this to evaluate

$$\langle jm|[J_\pm, Z_\alpha]|jm'\rangle = \begin{cases} \langle jm|(J_\pm Z_\alpha - Z_\alpha J_\pm)|jm'\rangle \\[2mm] \hbar\sqrt{(1 \mp \alpha)(2 \pm \alpha)}\langle jm|Z_{\alpha\pm 1}|jm'\rangle \end{cases} \qquad (5.2.18)$$

by the two alternatives indicated, where $|jm\rangle$ is the joint eigenket of $\vec{J}^2$ and J_z as usual. This establishes

$$\sqrt{(1 \mp \alpha)(2 \pm \alpha)}\langle jm|Z_{\alpha\pm 1}|jm'\rangle$$
$$+ \langle jm|Z_\alpha|jm' \pm 1\rangle\sqrt{(j \mp m')(j \pm m' + 1)}$$
$$= \sqrt{(j \pm m')(j \mp m' + 1)}\langle jm \mp 1|Z_\alpha|jm'\rangle. \quad (5.2.19)$$

We compare this recurrence relation with the recurrence relation obeyed by the Clebsch–Gordan coefficients $\langle j_1, j_2; jm | (j_1 m_1)(j_1 m_2) \rangle$ that we obtain by sandwiching $J_{1\pm} + J_{2\pm} = J_\pm$ with this bra-ket pair:

$$
\langle j_1, j_2; jm | (j_1 m_1 \pm 1)(j_1 m_2) \rangle \sqrt{(j_1 \mp m_1)(j_1 \pm m_1 + 1)}
$$
$$
+ \langle j_1, j_2; jm | (j_1 m_1)(j_1 m_2 \pm 1) \rangle \sqrt{(j_2 \mp m_2)(j_2 \pm m_2 + 1)}
$$
$$
= \sqrt{(j \pm m)(j \mp m + 1)} \langle j_1, j_2; jm \mp 1 | (j_1 m_1)(j_1 m_2) \rangle .
$$

$$(5.2.20)$$

This latter recurrence relation (5.2.20) turns into the former one in (5.2.19) by the replacements

$$
(j_1, m_1, j_2, m_2, j, m) \longrightarrow (1, \alpha, j, m', j, m) , \qquad (5.2.21)
$$

which is to say that we have the same recurrence relation twice. Since it is a linear relation, the solution of one recurrence relation is proportional to the solution of the other. This tells us that

$$
\langle jm | Z_\alpha | jm' \rangle = U_{\vec{S}, j} \langle 1j; jm | (1\alpha)(jm') \rangle \qquad (5.2.22)
$$

where $U_{\vec{S}, j}$ is a *universal* constant in the sense that it does not depend on m, m' or α but is specified by $\vec{S}$ and j solely. In particularly, we have the $\alpha = 0$ statement for $m = m'$,

$$
\langle jm | S_z | jm \rangle = U_{\vec{S}, j} \langle 1j; jm | (10)(jm) \rangle . \qquad (5.2.23)
$$

In the derivation of this result, what was important about $\vec{S}$ is its vector character. We can repeat the whole story for any other vector operator, for which $\vec{J}$ is the most particular example. Thus, we also have

$$
\langle jm | J_z | jm \rangle = U_{\vec{J}, j} \langle 1j; jm | (10)(jm) \rangle . \qquad (5.2.24)
$$

5-8   Write

$$
\vec{J} \cdot \vec{S} = J_z Z_0 - \frac{1}{\sqrt{2}} J_+ Z_{-1} - \frac{1}{\sqrt{2}} J_- Z_{+1}
$$

and use then an analogous argument to establish

$$
\langle jm | \vec{J} \cdot \vec{S} | jm \rangle = U_{\vec{S}, j} c_j ,
$$

where the c_j are coefficients that can be expressed in terms of Clebsch–Gordan coefficients.

5-9 Further establish that

$$\langle jm|\vec{J}^{2}|jm\rangle = U_{\bar{j},j}c_{j}$$

with the same coefficients c_j.

We combine these four statements into one:

$$\langle jm|S_z|jm\rangle = \frac{\langle jm|\vec{J}\cdot\vec{S}|jm\rangle}{\langle jm|\vec{J}^{2}|jm\rangle}\langle jm|J_z|jm\rangle. \qquad (5.2.25)$$

Of the three expectation values on the right-hand side, two are immediately available,

$$\langle jm|J_z|jm\rangle = \hbar m \,,$$

$$\langle jm|\vec{J}^{2}|jm\rangle = \hbar^{2}j(j+1)\,, \qquad (5.2.26)$$

and the third can be written as

$$\langle jm|\vec{J}\cdot\vec{S}|jm\rangle = \frac{1}{2}\langle jm|\left(\vec{J}^{2}-\vec{L}^{2}+\vec{S}^{2}\right)|jm\rangle$$

$$= \frac{1}{2}\hbar^{2}[j(j+1)-l(l+1)+s(s+1)] \qquad (5.2.27)$$

after taking a look at the square of $\vec{L} = \vec{J} - \vec{S}$. We remember, of course, that $|jm\rangle$ is here a shorthand notation for the joint eigenstates of $\vec{L}^{2}, \vec{S}^{2}, \vec{J}^{2}$ and J_z, denoted by $|l, s, j, m_j\rangle$ in (5.2.12).

The bottom line is thus

$$\langle jm|S_z|jm\rangle = \frac{j(j+1)-l(l+1)+s(s+1)}{2j(j+1)}\hbar m\,, \qquad (5.2.28)$$

which we combine with $\vec{L} + g\vec{S} = \vec{J} + (g-1)\vec{S}$ to arrive at ($m \to m_j$ now)

$$\delta E_{\text{magn}} = \left[1 + (g-1)\frac{j(j+1)-l(l+1)+s(s+1)}{2j(j+1)}\right]m_j\mu_{\text{B}}B\,. \qquad (5.2.29)$$

In particular, for $g = 2$ this gives

$$\delta E_{\text{magn}} = g_{\text{eff}}m_j\mu_{\text{B}}B \qquad (5.2.30)$$

with an effective g-factor of

$$g_{\text{eff}} = \frac{3j(j+1)-l(l+1)+s(s+1)}{2j(j+1)}\,. \qquad (5.2.31)$$

5-10 Do you get the expected values for g_{eff} in the cases of $l = 0$ or $s = 0$?

5-11 At (5.2.14), we noted that $\delta E_{\text{magn}} = \mu_{\text{B}}$ is fourfold degenerate. Can you confirm this by using (5.2.30)? Hint: Remember the intricacies of perturbation theory for degenerate states, discussed in Section 6.6 of *Simple Systems*.

One may get the impression that it is not worth the trouble to use the $\vec{J}^2, J_z$ states for an evaluation of the magnetic energy shift. But in the most important application to the *Zeeman effect* in atomic physics — the lifting of the degeneracy in a magnetic field, the counterpart of the electric-field Stark effect discussed in Section 6.7 of *Simple Systems* (These effects are named after Pieter Zeeman and Johannes Stark, respectively.) — there is an additional complication. It is the magnetic interaction between the magnetic moments of the electrons and the magnetic moments of their orbits created by the electric currents associated with the moving electron. These magnetic moments are respectively proportional to $\vec{S}$ and $\vec{L}$, so that we have an additional term in the Hamilton operator of the form

$$H_{\text{LS}} = V_{\text{LS}} \vec{L} \cdot \vec{S}, \qquad (5.2.32)$$

where V_{LS} is a position-dependent coupling strength.

If such a *spin-orbit coupling* term is present, and not negligible, we cannot use the eigenstates of $\vec{L}, \vec{S}, L_z, S_z$ as unperturbed states because $\vec{L} \cdot \vec{S}$ does not commute with L_z and S_z. But it commutes with $\vec{J}^2$ and J_z. Similarly to what we found for $\vec{J} \cdot \vec{S}$ above and used in (5.2.27), we have

$$\vec{L} \cdot \vec{S} = \frac{1}{2}\left(\vec{L} + \vec{S}\right)^2 - \frac{1}{2}\vec{L}^2 - \frac{1}{2}\vec{S}^2$$
$$= \frac{1}{2}\left(\vec{J}^2 - \vec{L}^2 - \vec{S}^2\right) \qquad (5.2.33)$$

and can evaluate the expectation value of H_{LS} in accordance with

$$\langle H_{\text{LS}} \rangle = \langle V_{\text{LS}} \rangle \frac{\hbar^2}{2}[j(j+1) - l(l+1) - s(s+1)], \qquad (5.2.34)$$

where the expectation value $\langle V_{\text{LS}} \rangle$ involves the spatial part of the wave function only.

Chapter 6

Indistinguishable Particles

6.1 Indistinguishability

At the atomic level, the physical entities, such as electrons, protons, atoms, or photons, possess no individuality. One electron is the same as any other electron in all respects — one speaks of *identical* or *indistinguishable* particles. This fundamental indistinguishability goes beyond the similarity that one can have at the level of classical physics. For example, when two billard balls bounce, someone not observing the collision may not be able to tell which ball came from the right and which from the left, but this information can be obtained. After all, we can follow the trajectory of any chosen billard ball and so can tell a ball from its lookalike.

Not so at the quantum level, where electrons, say, do not have trajectories that would be so well defined that one can follow them during a collision act. As a consequence, there is no physical significance of the labels that we use in the formalism for a number of electrons. The Hamilton operator for two electrons, for instance,

$$H = \frac{1}{2M}\left(\vec{P}_1^2 + \vec{P}_2^2\right) + \frac{e^2}{\left|\vec{R}_1 - \vec{R}_2\right|} \qquad (6.1.1)$$

does not change if the labels are interchanged, $H(1,2) \to H(2,1)$. This *invariance under permutation* of the labels of indistinguishable particles must be possessed by *any* operator corresponding to a physical observable, the Hamilton operator being one of them. Another example is the total angular momentum of the two electrons,

$$\vec{J} = \underbrace{\vec{R}_1 \times \vec{P}_1 + \vec{R}_2 \times \vec{P}_2}_{\substack{\text{orbital} \\ \text{angular momentum}}} + \underbrace{\vec{S}_1 + \vec{S}_2}_{\text{spin}} ; \qquad (6.1.2)$$

quite visibly the permutation $1 \leftrightarrow 2$ has no effect on $\vec{J}$ as a whole or on its orbital and spin constituents.

When we consider the center-of-mass position $\vec{R}_{CM}$ and the center-of-mass momentum $\vec{P}_{CM}$,

$$\vec{R}_{CM} = \frac{1}{2}\left(\vec{R}_1 + \vec{R}_2\right), \qquad \vec{P}_{CM} = \vec{P}_1 + \vec{P}_2, \qquad (6.1.3)$$

we note that these are symmetric under $1 \leftrightarrow 2$ as well. Not so, however, for the relative position

$$\vec{R} = \vec{R}_1 - \vec{R}_2 \qquad (6.1.4)$$

and the relative momentum

$$\vec{P} = \frac{1}{2}\left(\vec{P}_1 - \vec{P}_2\right), \qquad (6.1.5)$$

which change sign when the labels are interchanged. As a consequence, physical observables cannot be functions of $\vec{R}$ and $\vec{P}$ themselves, only of their squares or their product.

We check this by inserting

$$\left.\begin{array}{c}\vec{R}_1 \\ \vec{R}_2\end{array}\right\} = \vec{R}_{CM} \pm \frac{1}{2}\vec{R}, \qquad \left.\begin{array}{c}\vec{P}_1 \\ \vec{P}_2\end{array}\right\} = \frac{1}{2}\vec{P}_{CM} \pm \vec{P} \qquad (6.1.6)$$

into the Hamilton operator (6.1.1),

$$\begin{aligned}
H &= \frac{1}{2M}\left[\left(\frac{1}{2}\vec{P}_{CM} + \vec{P}\right)^2 + \left(\frac{1}{2}\vec{P}_{CM} - \vec{P}\right)^2\right] + \frac{e^2}{|\vec{R}|} \\
&= \frac{1}{4M}\vec{P}_{CM} + \frac{1}{M}\vec{P}^2 + \frac{e^2}{|\vec{R}|} \equiv H_{CM} + H_{rel} \qquad (6.1.7)
\end{aligned}$$

where both the Hamilton operator for the center-of-mass motion,

$$H_{CM} = \frac{1}{4M}\vec{P}_{CM}^2 = \frac{1}{4M}\left(\vec{P}_1 + \vec{P}_2\right)^2, \qquad (6.1.8)$$

and that for the relative motion,

$$H_{rel} = \frac{1}{M}\vec{P}^2 + \frac{e^2}{|\vec{R}|} = \frac{1}{4M}\left(\vec{P}_1 - \vec{P}_2\right)^2 + \frac{e^2}{|\vec{R}_1 - \vec{R}_2|}, \qquad (6.1.9)$$

are invariant under the interchange of labels 1 and 2, as they should be.

There is an additional lesson here, namely that the center-of-mass motion is force-free motion of a system with mass $2M$, which is hardly surprising, and that the relative motion is that of a system with mass $\frac{1}{2}M$ in the potential energy $e^2/|\vec{R}|$. This reduction of the more complicated two-particle problem to two effective single-particle problems is extremely useful because it enables us to solve the two-particle problem by solving the one-particle problem of H_{rel}. That relative motion is not the motion of a real physical object, it is the motion of an effective object, with an effective mass that is half the physical mass of either one of the two indistinguishable particles that are interacting physically.

Of course, this reduction is familiar from classical physics, and we recall that in the more general situation of two physical particles with different masses M_1, M_2, the effective mass of the relative motion is $(1/M_1 + 1/M_2)^{-1}$, which becomes $\frac{1}{2}M$ for $M_1 = M_2 = M$.

6-1 Write the orbital angular momentum

$$\vec{L} = \vec{R}_1 \times \vec{P}_1 + \vec{R}_2 \times \vec{P}_2$$

in terms of center-of-mass and relative operators. Do you get what you expect?

6.2 Bosons and fermions

So, having established that all physical observables are invariant under the permutation of the labels of indistinguishable particles, what about the kets, bras, Schrödinger wave functions, and so forth? We have

$$i\hbar\frac{\partial}{\partial t}\langle 1,2,t| = \langle 1,2,t|H(1,2) \qquad (6.2.1)$$

for one labeling, and

$$i\hbar\frac{\partial}{\partial t}\langle 2,1,t| = \langle 2,1,t|H(2,1) \qquad (6.2.2)$$

for the permuted labeling. But $H(1,2) = H(2,1)$, so that

$$i\hbar\frac{\partial}{\partial t}\langle 2,1,t| = \langle 2,1,t|H(1,2), \qquad (6.2.3)$$

which tells us that $\langle 1,2,t|$ and $\langle 2,1,t|$ are solutions of the same equation of motion. Since this must be generally true, not just for some particular situations, we infer that $\langle 2, 1, t|$ differs from $\langle 1, 2, t|$ only by an overall phase factor,

$$\langle 2, 1, t| = e^{i\varphi}\langle 1, 2, t|, \tag{6.2.4}$$

and this statement must be independent of the originally chosen labeling, because that has no physical significance. Therefore, we must also have

$$\langle 1, 2, t| = e^{i\varphi}\langle 2, 1, t| \tag{6.2.5}$$

with the *same* phase factor. The two statements are only consistent if

$$e^{2i\varphi} = 1, \tag{6.2.6}$$

which offers the options

$$e^{i\varphi} = +1 \quad \text{or} \quad e^{i\varphi} = -1. \tag{6.2.7}$$

Both are conceivable, as both are consistent with our general conclusions from the indistinguishability of the two indistinguishable particles under consideration. And, in fact, Nature does make use of both options.

There are two kinds of particles,

> *bosons* have $\langle 1, 2| = \langle 2, 1|$, that is: the bras, kets, wave functions, ... are *symmetric* in the labels and do not change when the labels are permuted;
>
> *fermions* have $\langle 1, 2| = -\langle 2, 1|$, that is: the bras, kets, wave functions, ... are *antisymmetric* in the labels and are multiplied by -1 when the labels are permuted.

$$(6.2.8)$$

They are named after Satyendranath Bose and Enrico Fermi. Their fundamental symmetry properties

$$\langle 1, 2| = \langle 2, 1| \text{ for bosons}, \quad \langle 1, 2| = -\langle 2, 1| \text{ for fermions} \tag{6.2.9}$$

are often referred to as *Bose–Einstein statistics* or *Fermi–Dirac statistics*, respectively, thereby honoring as well the contributions by Albert Einstein and Paul A. M. Dirac.

It is a well-established experimental fact, understood to a very large extent, but perhaps not fully, as a consequence of fundamental principles in quantum field theory, that all fermions have half-integer spin, and all bosons have integer spin:

$$\text{fermions have } s = \tfrac{1}{2}, \tfrac{3}{2}, \tfrac{5}{2}, \ldots,$$
$$\text{bosons have } s = 0, 1, 2, \ldots. \tag{6.2.10}$$

Clearly, this link of statistical properties with intrinsic angular momentum, famously known as Wolfgang Pauli's *spin-statistics theorem* is a fundamental property of quantum objects.

The basic building blocks of atoms and atomic nuclei — the electrons, protons, and neutrons — all are spin-$\frac{1}{2}$ particles and thus fermions. Most nuclei have an even number of neutrons, and since there are as many electrons as there are protons in a neutral atom, we observe that most neutral atoms are composed of an even number of fermions. Therefore, their total spin is integer, and the neutral atoms as a whole are bosons. Simply ionized atoms have one electron removed, so that they have an odd number of fermions as constituents, and thus have half-integer total spin, which implies that they are fermions. Likewise doubly ionized atoms are bosons, and so forth.

In the much rarer cases where the nucleus has an odd number of neutrons, matters are reversed: The neutral atom is a fermion, the singly ionized atom is a boson, the doubly ionized atom is a fermion, and so forth.

As an important example, let us consider a system composed of two electrons. Each electron has two spin states, "up" and "down", that we denote by $|\uparrow\rangle$ and $|\downarrow\rangle$ as we did in Section 4.2.1, where we established the two-particle states to total spin 0 and total spin 1. The spin part of the *singlet* state to $s = 0$,

$$\frac{1}{\sqrt{2}}\Big(|\uparrow\downarrow\rangle - |\downarrow\uparrow\rangle \Big) \tag{6.2.11}$$

is *antisymmetric* under the permutation of the electrons whereas the spin parts of the three *triplet* states,

$$|\uparrow\uparrow\rangle, \quad \frac{1}{\sqrt{2}}\Big(|\uparrow\downarrow\rangle + |\downarrow\uparrow\rangle \Big), \quad |\downarrow\downarrow\rangle \tag{6.2.12}$$

are *symmetric* under the permutation of the electrons. It follows that the *spatial wave function* $\psi(\vec{r}_1, \vec{r}_2)$ that goes with the singlet states must be symmetric,

$$\psi_{\text{singlet}}(\vec{r}_1, \vec{r}_2) = \psi_{\text{singlet}}(\vec{r}_2, \vec{r}_1), \tag{6.2.13}$$

but that for a triplet state must be antisymmetric,

$$\psi_{\text{triplet}}(\vec{r}_1, \vec{r}_2) = -\psi_{\text{triplet}}(\vec{r}_2, \vec{r}_1). \tag{6.2.14}$$

In this manner, the *total* state ket, the product of the spin part and the spatial part, is antisymmetric for both singlet and triplet.

A general construction begins with single-electron states $|a_1\rangle, |a_2\rangle, \ldots$ normalized and pairwise orthogonal, whereby the labels a_j comprise all quantum numbers, those for the spatial degrees of freedom and for the spin. Then

$$\left| \{a_j, a_k\} \right\rangle = \frac{1}{\sqrt{2}} \left(\left| a_j, a_k \right\rangle - \left| a_k, a_j \right\rangle \right) \quad \text{with} \quad j < k \tag{6.2.15}$$

is a corresponding set of two-electron states, whereby the restriction to $j < k$ avoids double counting. If we adopt a widespread but rather awkward notation,

$$\left| a_j, a_k \right\rangle = \left| a_j \right\rangle_1 \left| a_k \right\rangle_2, \tag{6.2.16}$$

we can write this as

$$\left| \{a_j, a_k\} \right\rangle = \frac{1}{\sqrt{2}} \det \begin{pmatrix} \left| a_j \right\rangle_1 & \left| a_k \right\rangle_1 \\ \left| a_j \right\rangle_2 & \left| a_k \right\rangle_2 \end{pmatrix}, \tag{6.2.17}$$

which is known as the *Slater determinant*, named after John Slater. Such products of kets are best understood as products of their numerical representatives, the corresponding wave functions, and then the determinant construction is very useful, because one can easily generalize it to three and more electrons. For example, for three electrons one would have

$$\left| \{a_j, a_k, a_l\} \right\rangle = \frac{1}{\sqrt{6}} \det \begin{pmatrix} \left| a_j \right\rangle_1 & \left| a_k \right\rangle_1 & \left| a_l \right\rangle_1 \\ \left| a_j \right\rangle_2 & \left| a_k \right\rangle_2 & \left| a_l \right\rangle_2 \\ \left| a_j \right\rangle_3 & \left| a_k \right\rangle_3 & \left| a_l \right\rangle_3 \end{pmatrix}$$

$$= \frac{1}{\sqrt{6}} \left(\left| a_j, a_k, a_l \right\rangle + \left| a_k, a_l, a_j \right\rangle + \left| a_l, a_j, a_k \right\rangle \right.$$

$$\left. - \left| a_j, a_l, a_k \right\rangle - \left| a_k, a_j, a_l \right\rangle - \left| a_l, a_k, a_j \right\rangle \right) \tag{6.2.18}$$

which is clearly antisymmetric under permutations of any pair of electrons, $1 \leftrightarrow 2$ or $2 \leftrightarrow 3$ or $3 \leftrightarrow 1$.

6-2   Consider the possible states of two electrons. What is the value of $\vec{S}_1 \cdot \vec{S}_2$ in the singlet state? What is it in the triplet state? Use this to express the projection operators for the singlet and triplet subspaces as simple functions of $\vec{S}_1 \cdot \vec{S}_2$.

6-3   Evaluate the expectation value $\langle \uparrow\downarrow | \vec{S}_1 \cdot \vec{S}_2 | \uparrow\downarrow \rangle$ and explain the physical significance of the number you get. Hint: What does the expectation value of $\vec{S}_1 \cdot \vec{S}_2$ tell you about the singlet fraction and the triplet fraction in the two-electron state under consideration?

6.3   Scattering of two indistinguishable particles

The symmetry of the wave function for two indistinguishable particles has quite a lot of phenomenological manifestations, of which the modification of the differential cross section for two-particle collisions is perhaps the simplest one. Consider thus the scattering of two electrons, which approach each other with equal but opposite momentum in the center-of-mass rest frame:

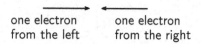

one electron   one electron
from the left   from the right

After the scattering we have escaping electrons

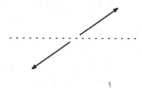

but we cannot tell which electron came from the left and which from the right. The two cases of deflection angle θ

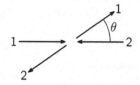

and of deflection angle $\pi - \theta$

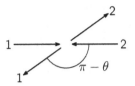

are utterly indistinguishable. Therefore, we must add the two respective scattering *amplitudes* to obtain the total scattering amplitude, before we square it to get the differential cross section,

$$\frac{\mathrm{d}\sigma}{\mathrm{d}\Omega}(\theta) = |f(\theta) \pm f(\pi - \theta)|^2, \qquad (6.3.1)$$

whereby the upper sign applies for a symmetric spatial wave function (singlet), and the lower sign applies when the spatial wave function is antisymmetric (triplet).

Of particular interest is the *right-angle scattering*, that is scattering under angle $\theta = \pi/2 = 90°$,

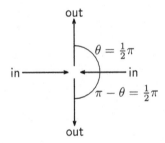

for which

$$\frac{\mathrm{d}\sigma}{\mathrm{d}\Omega}\left(\theta = \frac{\pi}{2}\right) = \left|f\left(\frac{\pi}{2}\right) \pm f\left(\frac{\pi}{2}\right)\right|^2 = \begin{cases} 4\left|f\left(\dfrac{\pi}{2}\right)\right|^2 & \text{for the singlet,} \\ 0 & \text{for the triplet.} \end{cases}$$
$$(6.3.2)$$

For example, if both electrons are spin-up, we have the spin ket $|\uparrow\uparrow\rangle$ which is symmetric, so no scattering under $\theta = 90°$ will be observed. By contrast, if the electron from the left is $\uparrow$, that from the right $\downarrow$, we have a spin ket

$$|\uparrow\downarrow\rangle = \frac{1}{\sqrt{2}}\Big[\underbrace{\frac{1}{\sqrt{2}}\big(|\uparrow\downarrow\rangle + |\downarrow\uparrow\rangle\big)}_{\text{triplet}} + \underbrace{\frac{1}{\sqrt{2}}\big(|\uparrow\downarrow\rangle - |\downarrow\uparrow\rangle\big)}_{\text{singlet}} \Big] \qquad (6.3.3)$$

which is an equal-weight superposition of the singlet state with a triplet state, so that we get 50% of each,

$$\frac{d\sigma}{d\Omega} = \frac{1}{2}|f(\theta) - f(\pi - \theta)|^2 + \frac{1}{2}|f(\theta) + f(\pi - \theta)|^2$$
$$= |f(\theta)|^2 + |f(\pi - \theta)|^2 . \qquad (6.3.4)$$

This is *as if* there were distinguishable particles, but we just do not know which is which. For $\theta = \pi/2$, the outgoing electrons after the scattering will be necessarily in the singlet state because the triplet state does not scatter under 90°.

If the electrons are unpolarized before the scattering, then all spin states are equally probable and we shall have the singlet in $1/4 = 25\%$ of all cases and the triplet for $3/4 = 75\%$. The observed differential cross section is then

$$\frac{d\sigma}{d\Omega} = \frac{1}{4}|f(\theta) + f(\pi - \theta)|^2 + \frac{3}{4}|f(\theta) - f(\pi - \theta)|^2$$
$$= |f(\theta)|^2 + |f(\pi - \theta)|^2 + \left(\frac{1}{4} - \frac{3}{4}\right)2\operatorname{Re}\big(f(\theta)^* f(\pi - \theta)\big)$$
$$= \underbrace{|f(\theta)|^2 + |f(\pi - \theta)|^2}_{\substack{\text{classically not} \\ \text{distinguished} \\ \text{particles}}} - \underbrace{\operatorname{Re}\big(f(\theta)^* f(\pi - \theta)\big)}_{\substack{\text{correction for} \\ \text{quantum} \\ \text{indistinguishability}}} , \qquad (6.3.5)$$

which has the terms of (6.3.4) for classically not distinguished particles plus a correction that accounts for the quantum-mechanical indistinguishability. At scattering angle $\theta = 90° = \pi/2$, we thus get

$$\frac{d\sigma}{d\Omega}\left(\theta = \frac{\pi}{2}\right) = 2\left|f\left(\frac{\pi}{2}\right)\right|^2 - \left|f\left(\frac{\pi}{2}\right)\right|^2$$
$$= \frac{1}{2} \times 2\left|f\left(\frac{\pi}{2}\right)\right|^2 , \qquad (6.3.6)$$

that is: a reduction by 50%.

So far these remarks apply to electrons, or more generally, to any pair of indistinguishable spin-$\frac{1}{2}$ particles. For a pair of indistinguishable particles with spin $s = 0, \frac{1}{2}, 1, \frac{3}{2}, 2, \frac{5}{2}, \dots$, we have the general relation (6.3.1) with its distinction between symmetric and antisymmetric spatial wave function. For bosons the spatial wave function has the same permutation symmetry as the spin part, and for fermions they have opposite permutation symmetry.

Now, two particles with spin s each have $(2s+1)^2$ spin states in total with magnetic quantum numbers m_1, m_2, say. There are the antisymmetric superpositions

$$\frac{1}{\sqrt{2}}\Big(\big|m_1, m_2\big\rangle - \big|m_2, m_1\big\rangle\Big) \quad \text{with} \quad m_1 \neq m_2, \tag{6.3.7}$$

which are $\frac{1}{2}\big[(2s+1)^2 - (2s+1)\big] = (2s+1)s$ in number, and there are the symmetric superpositions

$$\frac{1}{\sqrt{2}}\Big(\big|m_1, m_2\big\rangle + \big|m_2, m_1\big\rangle\Big) \quad \text{with} \quad m_1 \neq m_2$$

$$\text{and} \quad \big|m_1, m_1\big\rangle \quad \text{that is} \quad m_1 = m_2, \tag{6.3.8}$$

which are $\frac{1}{2}\big[(2s+1)^2 - (2s+1)\big] + (2s+1) = (2s+1)(s+1)$ in number. So, the fraction of symmetric spin states is $(s+1)/(2s+1)$ and that of the antisymmetric states is $s/(2s+1)$. Therefore, the fraction of *symmetric spatial states* is $(s+1)/(2s+1)$ for bosons and $s/(2s+1)$ for fermions, and the fraction of *antisymmetric spatial states* is $s/(2s+1)$ for bosons and $(s+1)/(2s+1)$ for fermions. In summary, this tells us that the differential cross section for an unpolarized pair of indistinguishable particles is

$$\frac{d\sigma}{d\Omega}(\theta) = \frac{s+1}{2s+1}\big|f(\theta) \pm f(\pi - \theta)\big|^2 + \frac{s}{2s+1}\big|f(\theta) \mp f(\pi - \theta)\big|^2$$

$$\text{for} \quad \begin{cases} \text{bosons:} & s = 0, 1, 2, \ldots \\ \text{fermions:} & s = \frac{1}{2}, \frac{3}{2}, \frac{5}{2}, \ldots \end{cases} \tag{6.3.9}$$

or

$$\frac{d\sigma}{d\Omega}(\theta) = \big|f(\theta)\big|^2 + \big|f(\pi - \theta)\big|^2 \pm \frac{2}{2s+1}\text{Re}\big(f(\theta)^* f(\pi - \theta)\big). \tag{6.3.10}$$

For $\theta = \pi/2$, this gives

$$\frac{d\sigma}{d\Omega}\Big(\frac{\pi}{2}\Big) = 2\Big|f\Big(\frac{\pi}{2}\Big)\Big|^2\Big(1 \pm \frac{1}{2s+1}\Big) \quad \text{for} \quad \begin{cases} \text{bosons} \\ \text{fermions} \end{cases} \tag{6.3.11}$$

where the cross section for classically not distinguished particles, $2\big|f\big(\frac{\pi}{2}\big)\big|^2$, is multiplied by

$$1 \pm \frac{1}{2s+1} = \begin{cases} 2, \frac{4}{3}, \frac{6}{5}, \frac{8}{7}, \ldots & \text{for} \quad s = 0, 1, 2, \ldots, \\ \frac{1}{2}, \frac{3}{4}, \frac{5}{6}, \frac{7}{8}, \ldots & \text{for} \quad s = \frac{1}{2}, \frac{3}{2}, \frac{5}{2}, \ldots. \end{cases} \tag{6.3.12}$$

There is an enhancement for bosons and a reduction for fermions. Note that the two sequences are reciprocals of each other.

6.4 Two-electron atoms

6.4.1 *Variational estimate for the ground state*

The fermion nature of electrons is of utter importance and great consequence for the structure of atoms. To get a first glimpse at this matter we consider two-electron atoms, such as neutral helium, singly-ionized lithium, doubly-ionized beryllium, ..., but also the negative hydrogen ion. All of these have two electrons bound to a nucleus, which we shall regard as an infinitely massive point charge of strength Ze, with

nuclear charge Z	atom
1	H^-
2	He
3	Li^+
4	Be^{++}

and so forth. For all of these, the Hamilton operator is

$$H = \frac{1}{2M}\left(\vec{P}_1^2 + \vec{P}_2^2\right) - \left(\frac{Ze^2}{|\vec{R}_1|} + \frac{Ze^2}{|\vec{R}_2|}\right) + \frac{e^2}{|\vec{R}_1 - \vec{R}_2|}$$
$$= H_{\text{kin}} + H_{\text{Ne}} + H_{\text{ee}}, \tag{6.4.1}$$

taking into account the kinetic energy of both electrons

$$H_{\text{kin}} = \frac{1}{2M}\left(\vec{P}_1^2 + \vec{P}_2^2\right), \tag{6.4.2}$$

their interaction energy with the nucleus

$$H_{\text{Ne}} = -Ze^2\left(\frac{1}{|\vec{R}_1|} + \frac{1}{|\vec{R}_2|}\right), \tag{6.4.3}$$

and the electrostatic interaction energy between the electrons

$$H_{\text{ee}} = \frac{e^2}{|\vec{R}_1 - \vec{R}_2|}. \tag{6.4.4}$$

We are neglecting some finer details including, in particular, all spin-dependent interactions such as the magnetic interaction between the magnetic moments carried by the electrons. Thus our approximate treatment has no interactions that depend on the absolute or relative orientation of the two electron spins.

Nevertheless there is a spin dependence, namely a dependence on the total spin $\vec{S} = \vec{S}_1 + \vec{S}_2$, because the singlet sector, characterized by $\vec{S}^2 = 0$ or $\vec{S}_1 \cdot \vec{S}_2 = -\frac{3}{4}\hbar^2$, has symmetric spatial wave functions,

$$\psi_{\text{singlet}}(\vec{r}_1, \vec{r}_2) = \psi_{\text{singlet}}(\vec{r}_2, \vec{r}_1), \qquad (6.4.5)$$

whereas the triplet sector, characterized by $\vec{S}^2 = 2\hbar^2$ or $\vec{S}_1 \cdot \vec{S}_2 = \frac{1}{4}\hbar^2$, has antisymmetric spatial wave functions,

$$\psi_{\text{triplet}}(\vec{r}_1, \vec{r}_2) = -\psi_{\text{triplet}}(\vec{r}_2, \vec{r}_1). \qquad (6.4.6)$$

Wave functions from these two classes are to be considered when searching for the eigenvalues of $H = H_{\text{kin}} + H_{\text{Ne}} + H_{\text{ee}}$ and so we should expect to have different sets of eigenvalues in the singlet and the triplet sectors.

Let us begin with the ground state. If it were not for the interaction energy, we would have just

$$H_{\text{kin}} + H_{\text{Ne}} = \sum_{j=1}^{2} \left(\frac{1}{2M} \vec{P}_j^2 - \frac{Ze^2}{|\vec{R}_j|} \right) \qquad (6.4.7)$$

which is the sum of two single-particle operators, each of the hydrogenic kind studied in Section 5.1 of *Simple Systems*. Accordingly, any product $\psi_1(\vec{r}_1)\psi_2(\vec{r}_2)$, of the wave function $\psi_1(\vec{r}_1)$ to an eigenstate of the $j = 1$ term and the wave function $\psi_2(\vec{r}_2)$ to an eigenstate of the $j = 2$ term, is the wave function to an eigenstate of $H_{\text{kin}} + H_{\text{Ne}}$. The lowest energy obtained this way is for the product

$$\psi_{1s}(\vec{r}_1)\psi_{1s}(\vec{r}_2) \qquad (6.4.8)$$

of the two $1s$-hydrogenic wave functions, for which

$$\langle (H_{\text{kin}} + H_{\text{Ne}}) \rangle = -2Z^2 \text{Ry} \qquad (6.4.9)$$

with $\text{Ry} = Me^4/(2\hbar^2) = 13.6\,\text{eV}$, the familiar Rydberg constant, the natural unit of atomic binding energies (Janne Rydberg). But this product is a symmetric wave function, so that the minimum of $H_{\text{kin}} + H_{\text{Ne}}$ is achieved for a singlet state. We can now treat the interaction energy H_{ee} as a perturbation and get a correction

$$\langle H_{\text{ee}} \rangle = \int (\mathrm{d}\vec{r}_1)(\mathrm{d}\vec{r}_2) \, |\psi_{1s}(\vec{r}_1)|^2 \frac{e^2}{|\vec{r}_1 - \vec{r}_2|} |\psi_{1s}(\vec{r}_2)|^2 \qquad (6.4.10)$$

to the $-2Z^2\,\text{Ry}$ of (6.4.9).

In the triplet sector, the simplest wave function that we can build for $H_{\text{kin}} + H_{\text{Ne}}$ is

$$\frac{1}{\sqrt{2}}\Big(\psi_{1s}(\vec{r}_1)\psi_{2s}(\vec{r}_2) - \psi_{2s}(\vec{r}_1)\psi_{1s}(\vec{r}_2)\Big) \tag{6.4.11}$$

(or ψ_{2p} in place of ψ_{2s}) which has one electron in the ground state of the hydrogenic single-electron Hamilton operator, and the other in one of the first excited states. Now

$$\big\langle(H_{\text{kin}} + H_{\text{Ne}})\big\rangle = -\left(Z^2 + \frac{Z^2}{4}\right)\text{Ry} = -\frac{5}{4}Z^2\,\text{Ry} \tag{6.4.12}$$

which exceeds $-2Z^2\,\text{Ry}$ by much more than $\langle H_{\text{ee}}\rangle$ can possibly account for. We conclude that the triplet state with lowest energy is an excited state; the ground state is a spin singlet.

The hydrogenic ground-state wave function

$$\psi_{1s}(\vec{r}) = \sqrt{\frac{Z^3}{\pi a_0^3}}\,\mathrm{e}^{-Zr/a_0}\,, \tag{6.4.13}$$

with the Bohr radius $a_0 = \hbar^2/(Me^2) = 0.529\,\text{Å}$, is suitable for an electron exposed to the nucleus all by itself. But now we have a second electron nearby, and this suggests to account for the electrostatic shielding of the second electron in a simple manner, namely by using

$$\psi_\kappa(\vec{r}) = \sqrt{\kappa^3/\pi}\,\mathrm{e}^{-\kappa r} \tag{6.4.14}$$

instead, with an adjustable value for κ. Eventually, we expect to have $\kappa = Z_{\text{eff}}/a_0$ for a good choice whereby the effective value Z_{eff} for the nuclear charge should be between Z (no shielding) and $Z - 1$ (full shielding of one charge unit),

$$Z - 1 < Z_{\text{eff}} < Z \quad \text{(expected)}. \tag{6.4.15}$$

Applying the Rayleigh–Ritz variational method of Section 6.3 in *Simple Systems*, we choose the optimal value of κ by requiring that

$$\psi_{\text{singlet}}(\vec{r}_1, \vec{r}_2) = \psi_\kappa(\vec{r}_1)\psi_\kappa(\vec{r}_2) \tag{6.4.16}$$

gives the minimal expectation value $\langle H\rangle$ of the Hamilton operator (6.4.1) for this single-parameter class of wave functions. The single-electron energies are

$$\langle(H_{\text{kin}} + H_{\text{Ne}})\rangle_\kappa = 2 \int (\mathrm{d}\vec{r}) \left(\frac{\hbar^2}{2M} \left| \vec{\nabla}\psi_\kappa(\vec{r}) \right|^2 - \frac{Ze^2}{r} |\psi_\kappa(\vec{r})|^2 \right), \quad (6.4.17)$$

where the prefactor just doubles the contribution from one electron that is spelt out explicitly. The kinetic energy therein involves the integral

$$\frac{M}{\hbar^2}\langle H_{\text{kin}} \rangle = \int (\mathrm{d}\vec{r}) \left| \vec{\nabla}\psi_\kappa(\vec{r}) \right|^2 = \int (\mathrm{d}\vec{r}) \frac{\kappa^3}{\pi} \left| -\kappa \frac{\vec{r}}{r} \mathrm{e}^{-\kappa r} \right|^2$$

$$= 4\pi \int_0^\infty \mathrm{d}r\, r^2 \frac{\kappa^5}{\pi} \mathrm{e}^{-2\kappa r} = 4\pi \frac{\kappa^5}{\pi} \frac{2!}{(2\kappa)^3} = \kappa^2, \quad (6.4.18)$$

where we have an application of Leonhard Euler's factorial integral,

$$\int_0^\infty \mathrm{d}x\, x^n\, \mathrm{e}^{-x} = n!\,, \quad (6.4.19)$$

which makes another appearance in the nucleus-electron energy,

$$-\frac{1}{2Ze^2}\langle H_{\text{Ne}} \rangle = \int (\mathrm{d}\vec{r}) \frac{1}{r} |\psi_\kappa(\vec{r})|^2 = \int (\mathrm{d}\vec{r}) \frac{1}{r} \frac{\kappa^3}{\pi} \left| \mathrm{e}^{-\kappa r} \right|^2$$

$$= 4\pi \int_0^\infty \mathrm{d}r\, r \frac{\kappa^3}{\pi} \mathrm{e}^{-2\kappa r} = 4\pi \frac{\kappa^3}{\pi} \frac{1!}{(2\kappa)^2} = \kappa. \quad (6.4.20)$$

Taken together, the total single-particle energy is

$$\langle(H_{\text{kin}} + H_{\text{Ne}})\rangle_\kappa = \frac{(\hbar\kappa)^2}{M} - 2Ze^2\kappa. \quad (6.4.21)$$

Upon recalling that $e^2/a_0 = (\hbar/a_0)^2/M = 2\mathrm{Ry}$, we can also present this as

$$\langle(H_{\text{kin}} + H_{\text{Ne}})\rangle_\kappa = 2\big((\kappa a_0)^2 - 2Z\kappa a_0\big)\,\mathrm{Ry} \quad (6.4.22)$$

which acquires its minimal value of $-2Z^2\,\mathrm{Ry}$ for $\kappa a_0 = Z$, as it should.

The expectation value of the electron-electron interaction energy,

$$\langle H_{\text{ee}} \rangle_\kappa = \int (\mathrm{d}\vec{r}_1)(\mathrm{d}\vec{r}_2) \frac{e^2}{|\vec{r}_1 - \vec{r}_2|} |\psi_\kappa(\vec{r}_1)\psi_\kappa(\vec{r}_2)|^2$$

$$= e^2 \left(\frac{\kappa^3}{\pi} \right)^2 \int (\mathrm{d}\vec{r}_1)\, \mathrm{e}^{-2\kappa r_1} \int (\mathrm{d}\vec{r}_2) \frac{\mathrm{e}^{-2\kappa r_2}}{|\vec{r}_1 - \vec{r}_2|}, \quad (6.4.23)$$

requires the evaluation of this double integral. We first carry out the $(\mathrm{d}\vec{r}_2)$ integration, for which $\vec{r}_1$ has a fixed direction and length, which we take to be along the polar axis for the spherical coordinates in $\vec{r}_2$ space that we employ in

$$
\begin{aligned}
\int (d\vec{r}_2) \frac{e^{-2\kappa r_2}}{|\vec{r}_1 - \vec{r}_2|} &= \int_0^\infty dr_2\, r_2^2\, e^{-2\kappa r_2} 2\pi \int_0^\pi d\vartheta \frac{\sin \vartheta}{\sqrt{r_1^2 + r_2^2 - 2r_1 r_2 \cos \vartheta}} \\
&= \int_0^\infty dr_2\, r_2^2\, e^{-2\kappa r_2} 2\pi \frac{1}{r_1 r_2} \left. \sqrt{r_1^2 + r_2^2 - 2r_1 r_2 \cos \vartheta}\,\right|_{\vartheta=0}^\pi \\
&= \int_0^\infty dr_2\, r_2^2\, e^{-2\kappa r_2} \frac{2\pi}{r_1 r_2} (r_1 + r_2 - |r_1 - r_2|) \\
&= \int_0^\infty dr_2\, r_2^2\, e^{-2\kappa r_2} \frac{4\pi}{\mathrm{Max}\{r_1, r_2\}} .
\end{aligned}
\tag{6.4.24}
$$

Accordingly,

$$
\begin{aligned}
\langle H_{\mathrm{ee}} \rangle_\kappa = e^2 \left(\frac{\kappa^3}{\pi} \right)^2 (4\pi)^2 \int_0^\infty dr_1\, r_1^2\, e^{-2\kappa r_1} \\
\times \int_0^\infty dr_2\, r_2^2\, e^{-2\kappa r_2} \frac{1}{\mathrm{Max}\{r_1, r_2\}}
\end{aligned}
\tag{6.4.25}
$$

where the integral is twice what we get for $\mathrm{Max}\{r_1, r_2\} = r_2$ because the $\mathrm{Max}\{r_1, r_2\} = r_1$ part is the same. So,

$$
\begin{aligned}
\langle H_{\mathrm{ee}} \rangle &= e^2 (4\kappa^3)^2 2 \int_0^\infty dr_1\, r_1^2\, e^{-2\kappa r_1} \int_{r_1}^\infty dr_2\, r_2\, e^{-2\kappa r_2} \\
&= e^2 (4\kappa^3)^2 2 \int_0^\infty dr_1\, r_1^2\, e^{-2\kappa r_1} \left(r_1 \frac{e^{-2\kappa r_1}}{2\kappa} + \frac{e^{-2\kappa r_1}}{(2\kappa)^2} \right) \\
&= e^2 (4\kappa^3)^2 2 \left(\frac{3!}{2\kappa (4\kappa)^4} + \frac{2!}{(2\kappa)^2 (4\kappa)^3} \right) \\
&= \frac{5}{8} e^2 \kappa = \frac{5}{4} \kappa a_0 \,\mathrm{Ry} .
\end{aligned}
\tag{6.4.26}
$$

In summary, then,

$$
\begin{aligned}
\frac{1}{2\mathrm{Ry}} \langle H \rangle_\kappa &= \frac{1}{2\mathrm{Ry}} \langle (H_{\mathrm{kin}} + H_{\mathrm{Ne}} + H_{\mathrm{ee}}) \rangle_\kappa \\
&= (\kappa a_0)^2 - 2Z\kappa a_0 + \frac{5}{8}\kappa a_0 \\
&= \left[\kappa a_0 - \left(Z - \frac{5}{16} \right) \right]^2 - \left(Z - \frac{5}{16} \right)^2 \\
&\geq -\left(Z - \frac{5}{16} \right)^2 ,
\end{aligned}
\tag{6.4.27}
$$

so that the optimal choice for κ is

$$\kappa = Z_{\text{eff}}/a_0 \qquad (6.4.28)$$

with the effective nuclear charge given by

$$Z_{\text{eff}} = Z - \frac{5}{16}. \qquad (6.4.29)$$

The resulting optimized upper bound for the ground-state energy is

$$\langle H \rangle \leq -2Z_{\text{eff}}^2 \, \text{Ry} \qquad (6.4.30)$$

which is a lower bound on the binding energy,

$$-\langle H \rangle \geq 2Z_{\text{eff}}^2 \, \text{Ry} . \qquad (6.4.31)$$

Here is a comparison with experimental values for $-\langle H \rangle /(2\text{Ry})$:

Z	experiment	estimate	error(%)
1	0.52776	0.473	10.4
2	2.9038	2.848	2.0
3	7.2804	7.223	0.79
4	13.657	13.598	0.44

Thus, except for $Z = 1$, the negative hydrogen ion H^-, we fare very well with this rather simple estimate.

For H^- the lower bound of (6.4.31) is indeed so bad, that the estimated binding energy is *less* than the binding energy of $1\,\text{Ry} = \frac{1}{2} \times (2\,\text{Ry})$ for a single electron in neutral hydrogen. That is: for our estimate it is energetically favorable to have the second electron far away, which is to say there could not be a stable H^- ion. But in fact it can be made in copious amounts rather easily. It is clear, then, that a better estimate is needed to deal with H^-.

Perhaps the simplest better trial wave function is

$$\psi(\vec{r}_1, \vec{r}_2) \propto \psi_{\kappa_1}(r_1)\psi_{\kappa_2}(r_2) + \psi_{\kappa_2}(r_1)\psi_{\kappa_1}(r_2)$$
$$\propto e^{-(\kappa_1 r_1 + \kappa_2 r_2)} + e^{-(\kappa_2 r_1 + \kappa_1 r_2)}, \qquad (6.4.32)$$

which has two parameters, κ_1 and κ_2, rather than the single κ parameter of (6.4.16) and (6.4.14). This incorporates the simple idea that one electron will be closer to the nucleus and the other farther away, with the nuclear charge shielded more effectively for the far-away electron than the near-by one. The calculation is much more tedious for this wave function than for the $\kappa_1 = \kappa_2 = \kappa$ case above, and it is hardly worth the trouble for

$Z = 2, 3, 4, \ldots$, but there is a substantial improvement for $Z = 1$, inasmuch as we get

$$- \langle H \rangle \geq 0.513 \times 2\,\mathrm{Ry} \qquad (6.4.33)$$

which *exceeds* $\frac{1}{2} \times 2\,\mathrm{Ry}$ and thus confirms the existence of a stable H^- ion. This estimate is also quite good quantitatively, as the error is just 3%.

6.4.2 Perturbative estimate for the first excited state

For the first excited states, we shall be content with what we can learn from a perturbation-theoretical estimate. As discussed above, we use the unperturbed wave functions

$$\frac{1}{\sqrt{2}} [\psi_{1s}(\vec{r}_1)\psi_{2s}(\vec{r}_2) \pm \psi_{1s}(\vec{r}_2)\psi_{2s}(\vec{r}_1)] = \psi_{\pm}(\vec{r}_1, \vec{r}_2) \qquad (6.4.34)$$

for the singlet (upper sign) and the triplet (lower sign), respectively. The single-particle part of the Hamilton operator has these wave functions as eigenfunctions, so that we get [cf. (6.4.12)]

$$\langle (H_{\mathrm{kin}} + H_{\mathrm{Ne}}) \rangle = -\left(1 + \frac{1}{4}\right) Z^2\,\mathrm{Ry} = -\frac{5}{4} Z^2\,\mathrm{Ry}, \qquad (6.4.35)$$

as we have one electron in the hydrogenic $1s$ state (energy $= -Z^2\,\mathrm{Ry}$) and the other electron in the hydrogenic $2s$ state (energy $= -\frac{1}{4}Z^2\,\mathrm{Ry}$).

When evaluating the electron-electron interaction energy, which is the perturbation to the unperturbed single-electron part,

$$\langle H_{\mathrm{ee}} \rangle = \int (\mathrm{d}\vec{r}_1)(\mathrm{d}\vec{r}_2)\, \frac{e^2}{|\vec{r}_1 - \vec{r}_2|} \left| \psi_{\pm}(\vec{r}_1, \vec{r}_2) \right|^2, \qquad (6.4.36)$$

we encounter

$$\left| \psi_{\pm}(\vec{r}_1, \vec{r}_2) \right|^2 = \psi_{\pm}(\vec{r}_1, \vec{r}_2)^* \psi_{\pm}(\vec{r}_1, \vec{r}_2)$$

$$= \frac{1}{2} |\psi_{1s}(\vec{r}_1)|^2 |\psi_{2s}(\vec{r}_1)|^2 + \frac{1}{2} |\psi_{1s}(\vec{r}_2)|^2 |\psi_{2s}(\vec{r}_1)|^2$$

$$\pm \frac{1}{2} \psi_{1s}(\vec{r}_1)^* \psi_{2s}(\vec{r}_2)^* \psi_{1s}(\vec{r}_2) \psi_{2s}(\vec{r}_1)$$

$$\pm \frac{1}{2} \psi_{1s}(\vec{r}_2)^* \psi_{2s}(\vec{r}_1)^* \psi_{1s}(\vec{r}_1) \psi_{2s}(\vec{r}_2). \qquad (6.4.37)$$

Of these four terms, the first two terms contribute equally, and so do the second two terms. We adopt the convention that the argument of ψ_{1s}^* is $\vec{r}$

and that of ψ_{2s}^* is $\vec{r}'$ and so arrive at

$$\langle H_{ee} \rangle = \int (d\vec{r})(d\vec{r}')\, |\psi_{1s}(\vec{r})|^2 \frac{e^2}{|\vec{r}-\vec{r}'|} |\psi_{2s}(\vec{r}')|^2$$

$$\pm \int (d\vec{r})(d\vec{r}')\, \psi_{1s}^*(\vec{r})\psi_{2s}(\vec{r}) \frac{e^2}{|\vec{r}-\vec{r}'|} \psi_{2s}(\vec{r}')^*\psi_{1s}(\vec{r}')$$

$$= (\text{direct term}) \pm (\text{exchange term}).\qquad(6.4.38)$$

The *direct term* has the appearance of a classical electrostatic interaction energy between two charge distributions with the charge densities

$$-e|\psi_{1s}(\vec{r})|^2 \quad \text{and} \quad -e|\psi_{2s}(\vec{r}')|^2 \qquad (6.4.39)$$

as if the two electrons were smeared-out negative unit charges. No such interpretation can be given to the *exchange term*, it is of a purely quantum mechanical character. The exchange term has products like

$$\psi_{1s}(\vec{r})^*\psi_{2s}(\vec{r}) \qquad (6.4.40)$$

of orthogonal wave functions, which means that such a product cannot have a definite sign (even if we manage to make it real everywhere), and therefore the exchange term is usually much smaller than the direct term, of the order of $\frac{1}{100}$ Ry for helium. One can show, but we will not take the trouble, that the exchange term is positive. Of course, the direct term is immediately seen to be positive.

We so conclude that the $1s2s$ singlet state of He has a slightly larger energy than the $1s2s$ triplet state, the difference being twice the exchange term. The lowest-energy excited state is therefore the $1s2s$ triplet state, and it can only decay to the $(1s)^2$ ground state. But this decay process involves the change of the spin state from triplet to singlet, which means that a spin-flipping interaction is needed. The coupling to the electron spin is through its magnetic moment, so that we need a magnetic interaction to induce the $1s2s \to (1s)^2$, triplet $\to$ singlet transition. Such magnetic couplings to the radiation field are quite weak, and therefore the $1s2s$ triplet state is extremely long-lived. It is the prime example of a *metastable* state in an atom.

6-4 Evaluate the integrals for the direct and the exchange terms above, that is for $1s2s$, and state the energy difference between the singlet and triplet states in Rydberg units.

6.4.3   Self-consistent single-electron wave functions

Rather than using the one-parameter trial wave function (6.4.8), we could write, more generally

$$\psi(\vec{r}_1, \vec{r}_2) = \psi_0(\vec{r}_1)\psi_0(\vec{r}_2) \qquad (6.4.41)$$

with a normalized single-electron wave function $\psi_0(\vec{r})$,

$$\int (\mathrm{d}\vec{r}) \, |\psi_0(\vec{r})|^2 = 1, \qquad (6.4.42)$$

about whose structure we make no assumptions at the outset. The single-electron terms $H_{\mathrm{kin}} + H_{\mathrm{Ne}}$ then have expectation values that are just twice the contribution of one electron,

$$\langle (H_{\mathrm{kin}} + H_{\mathrm{Ne}}) \rangle = 2 \int (\mathrm{d}\vec{r}) \left[\frac{\hbar^2}{2M} \left| \vec{\nabla}\psi_0(\vec{r}) \right|^2 - \frac{Ze^2}{r} |\psi_0(\vec{r})|^2 \right], \quad (6.4.43)$$

and the electron-electron interaction term has the structure of a "direct term",

$$\langle H_{\mathrm{ee}} \rangle = \int (\mathrm{d}\vec{r})(\mathrm{d}\vec{r}') \, |\psi_0(\vec{r})|^2 \frac{e^2}{|\vec{r} - \vec{r}'|} |\psi_0(\vec{r}')|^2. \qquad (6.4.44)$$

As another application of the Rayleigh–Ritz method, we determine the best choice for $\psi_0(\vec{r})$ as the one that minimizes

$$\langle H \rangle = \langle (H_{\mathrm{kin}} + H_{\mathrm{Ne}} + H_{\mathrm{ee}}) \rangle, \qquad (6.4.45)$$

so we find it by requiring that $\delta\langle H \rangle = 0$ for variations of $\psi_0(\vec{r})$. These variations are subject to the normalization constraint, so that any permissible $\delta\psi_0(\vec{r})$ obeys

$$\int (\mathrm{d}\vec{r}) \left[\delta\psi_0(\vec{r})^* \psi_0(\vec{r}) + \psi_0(\vec{r})^* \delta\psi_0(\vec{r}) \right] = 0. \qquad (6.4.46)$$

For the response of $\langle H \rangle$ to variations of ψ_0 we get

$$\delta\langle H \rangle = 2 \int (\mathrm{d}\vec{r}) \left(\delta\psi_0^* \phi + \phi^* \delta\psi_0 \right) \qquad (6.4.47)$$

with

$$\phi(\vec{r}) = -\frac{\hbar^2}{2M} \vec{\nabla}^2 \psi_0(\vec{r}) - \frac{Ze^2}{r} \psi_0(\vec{r})$$
$$+ \int (\mathrm{d}\vec{r}') \frac{e^2}{|\vec{r} - \vec{r}'|} |\psi_0(\vec{r}')|^2 \psi_0(\vec{r}). \qquad (6.4.48)$$

Therefore, the variation of $\langle H \rangle$ vanishes if

$$\phi(\vec{r}) = \mathcal{E}\psi_0(\vec{r}) \tag{6.4.49}$$

with some real $\mathcal{E}$, the Lagrange parameter of the normalization constraint, named after Joseph L. Lagrange.

The equation that determines $\psi_0(\vec{r})$,

$$\left[-\frac{\hbar^2}{2M}\vec{\nabla}^2 + V(\vec{r}) \right]\psi_0(\vec{r}) = \mathcal{E}\psi_0(\vec{r}), \tag{6.4.50}$$

has the appearance of a single-particle Schrödinger eigenvalue equation, but this resemblance is somewhat misleading, inasmuch as the *effective potential*

$$V(\vec{r}) = -\frac{Ze^2}{r} + \int (\mathrm{d}\vec{r}') \frac{e^2}{|\vec{r} - \vec{r}'|} \left| \psi_0(\vec{r}') \right|^2 \tag{6.4.51}$$

contains the electrostatic potential of one electron and so involves the unknown wave function $\psi_0(\vec{r})$. In short, the equation that determines $\psi_0(\vec{r})$ is *nonlinear*. As a consequence, we cannot solve it with the methods developed for the linear Schrödinger equation. The crucial difference is that the equation for $\psi_0(\vec{r})$ does not obey the superposition principle: a linear combination of two different solutions is *not* a new solution.

The usual approach employs an iteration procedure:

(1) Take your present guess for $\psi_0(\vec{r})$ and evaluate $V(\vec{r})$ for it.

(2) Then solve (6.4.50) with this *fixed* $V(\vec{r})$, thereby obtaining an improved guess for $\psi_0(\vec{r})$ and an improved value for $\mathcal{E}$. $\tag{6.4.52}$

(3) Repeat as often as necessary.

In the present case where the external potential is spherically symmetric, the solution for $\psi_0(\vec{r})$ will also be spherically symmetric, which reduces the complexity of the numerical problem quite a bit.

6-5 Write $\psi_0(\vec{r}) = \frac{1}{r}u_0(r)$ and state the nonlinear differential eigenvalue equation obeyed by $u_0(r)$.

6-6 How is the expectation value $\langle H \rangle$ related to $\mathcal{E}$, once the solution $\psi_0(\vec{r})$ has been determined?

For an analogous treatment of the lowest-energy triplet state we need an antisymmetric trial function, for which

$$\psi(\vec{r}_1, \vec{r}_2) = \frac{1}{\sqrt{2}}[\psi_1(\vec{r}_1)\psi_2(\vec{r}_2) - \psi_2(\vec{r}_1)\psi_1(\vec{r}_2)] \qquad (6.4.53)$$

with

$$\int (\mathrm{d}\vec{r}) \, \psi_j(\vec{r})^* \psi_k(\vec{r}) = \delta_{jk} \quad \text{for} \quad j, k = 1, 2 \qquad (6.4.54)$$

is the simplest choice. The minimization of $\langle H \rangle$ under the constraint of this orthonormality condition, then gives a coupled system of equations of the basic structure that we have seen for $\psi_0(\vec{r})$ in (6.4.50). There is now one equation for $\psi_1(\vec{r})$ with an effective potential $V_1(\vec{r})$, and another equation for $\psi_2(\vec{r})$ with an effective potential $V_2(\vec{r})$. The three constraints (normalization of ψ_1, of ψ_2, and their orthogonality) require three Lagrange parameters, three numbers of the kind of the $\mathcal{E}$ in (6.4.49). Once the equations are set up, one solves them numerically by an iteration such as the one described in (6.4.52), except that we now have to determine two wave functions and three Lagrange parameters.

6-7 Derive these coupled equations for $\psi_1(\vec{r})$ and $\psi_2(\vec{r})$. Then show that you can reduce the number of Lagrange parameters from three to two by systematically ensuring that ψ_1 and ψ_2 are orthogonal.

6.5 A glimpse at many-electron atoms

This method of determining a best simple-structure approximation for the wave function can be applied to atoms with many electrons as well. But we need to be somewhat more systematic about the handling of the electron spin degrees of freedom. The wave function of a single electron is a two-component object ("spin-up" and "spin-down"), that of an electron pair has four components (one singlet state, three triplet states), and for N electrons we will have 2^N components to the wave function. Denoting by

$$\psi(\vec{r}_1, s_1; \vec{r}_2, s_2; \ldots; \vec{r}_N, s_N) \qquad (6.5.1)$$

the component that has spin labels $s_1, s_2, \ldots, s_N$ for the electrons located at $\vec{r}_1, \vec{r}_2, \ldots, \vec{r}_N$, respectively, we note that this wave function must change sign when any two labels are interchanged, such as

$$\psi\left(\ldots;\vec{r}_3,s_3;\ldots;\vec{r}_8,s_8;\ldots\right) = -\psi\left(\ldots;\vec{r}_8,s_8;\ldots;\vec{r}_3,s_3;\ldots\right). \qquad (6.5.2)$$

The simplest wave function with this property is given by a Slater determinant, such as those in (6.2.17) and (6.2.18) for two and three electrons, respectively. For N electrons we would write

$$\psi(\vec{r}_1,s_1;\vec{r}_2,s_2;\ldots;\vec{r}_N,s_N) \propto \det \begin{pmatrix} \psi_1(\vec{r}_1,s_1) & \psi_1(\vec{r}_2,s_2) & \cdots & \psi_1(\vec{r}_N,s_N) \\ \psi_1(\vec{r}_1,s_1) & \psi_2(\vec{r}_2,s_2) & \cdots & \psi_2(\vec{r}_N,s_N) \\ \vdots & \vdots & & \vdots \\ \psi_N(\vec{r}_1,s_1) & \psi_N(\vec{r}_2,s_2) & \cdots & \psi_N(\vec{r}_N,s_N) \end{pmatrix}$$
$$(6.5.3)$$

with

$$\int (\mathrm{d}\vec{r})\psi_j(\vec{r},s)^*\psi_k(\vec{r},s) = \delta_{jk}\,, \qquad (6.5.4)$$

that is: the spatial wave functions for equal spin labels must be orthogonal. For opposite spin labels, the orthogonality is enforced by the spin degree of freedom and the spatial wave function can be identical, as is the situation for $\psi(\vec{r}_1,\vec{r}_2)$ in (6.4.41).

When minimizing the expectation value of the multi-electron Hamilton operator for a wave function of the Slater-determinant form (6.5.3), we get a coupled system of equations for the wave functions $\psi_j(\vec{r},s)$, the *Hartree–Fock equations* which are named after Douglas R. Hartree and Vladimir A. Fock. These equations are solved numerically by an iteration that is analogous to the one described in (6.4.52). One gets quite reasonable estimates for ground-state energies of atoms, but finer details very often require better wave functions. Those are systematically available as weighted sums of several Slater determinants, for example. Further details are beyond the scope of these lectures.

Nevertheless we can gain a basic understanding of general properties of many-electron atoms by adopting a different point of view. Atoms are held together by the strong attractive force between the nuclear charge and the electron charges, which more than balances the repulsive forces between the electron charges. This suggests a brutal approximation in which we ignore the repulsion among the electrons and focus solely on the attraction by the nuclear charge. Then the electrons are individually obeying the Schrödinger equation for hydrogenic atoms, see (5.1.2) in *Simple Systems*, for which

$$-E_n = \frac{Z^2 e^2 / a_0}{2n^2} \tag{6.5.5}$$

is the single-electron binding energy in the nth Bohr shell ($n = 1, 2, 3, \ldots$) for nuclear charge Ze; and a_0 is the Bohr radius, $e^2/a_0 = 27.2\,\text{eV} = 2\,\text{Ry}$. A Bohr shell with principal quantum number n has n^2 orbital states, as we recall from the degeneracy discussion in Section 5.1 of *Simple Systems*, and since there are two spin states for each orbital state, there are in total $2n^2$ states in the nth Bohr radius.

Now, imagine that we have a total of n_s filled Bohr shells. Then the total binding energy is

$$-E = \sum_{n=1}^{n_s} 2n^2 (-E_n) = \frac{Z^2 e^2}{a_0} n_s \tag{6.5.6}$$

and the count of electrons is

$$N = \sum_{n=1}^{n_s} 2n^2 = \frac{2}{3}\left(n_s + \frac{1}{2}\right)^3 - \frac{1}{6}\left(n_s + \frac{1}{2}\right). \tag{6.5.7}$$

We solve the latter for n_s in terms of N by iterating

$$\begin{aligned}
n_s &= \left[\frac{3}{2}N + \frac{1}{4}\left(n_s + \frac{1}{2}\right)\right]^{1/3} - \frac{1}{2} \\
&= \left(\frac{3}{2}N\right)^{1/3}\left[1 + \frac{n_s + 1/2}{6N}\right]^{1/3} - \frac{1}{2},
\end{aligned} \tag{6.5.8}$$

that is: we begin with the 1st-order approximation

$$n_s \cong \left(\frac{3}{2}N\right)^{1/3} - \frac{1}{2}, \tag{6.5.9}$$

then insert this on the right-hand side of (6.5.8) and expand $[1 + \cdots]^{1/3}$ to 1st-order to arrive at the 2nd-order approximation

$$n_s \cong \left(\frac{3}{2}N\right)^{1/3}\left[1 + \frac{1}{18N}\left(\frac{3}{2}N\right)^{1/3}\right] - \frac{1}{2} \tag{6.5.10}$$

or

$$n_s = \left(\frac{3}{2}N\right)^{1/3} - \frac{1}{2} + \frac{1}{12}\left(\frac{3}{2}N\right)^{-1/3} + \cdots \tag{6.5.11}$$

where the ellipses stands for terms of order $\left(\frac{3}{2}N\right)^{-3/3}$, $\left(\frac{3}{2}N\right)^{-5/3}$, and so forth.

This 2nd-order approximation is fully satisfactory for the present purpose. It gives us

$$\frac{-E}{\frac{1}{2}Z^2e^2/a_0} = 2\left(\frac{3}{2}N\right)^{1/3} - 1 + \frac{1}{6}\left(\frac{3}{2}N\right)^{-1/3}$$
$$= 2.289N^{1/3} - 1 + 0.146N^{-1/3} \qquad (6.5.12)$$

for the energy of an atom with nuclear charge Ze and N electrons that do not interact with each other. In particular, we have for neutral atoms, $N = Z$,

$$\frac{-E}{\frac{1}{2}Z^2e^2/a_0} = 2.289Z^{1/3} - 1 + 0.146Z^{-1/3}. \qquad (6.5.13)$$

Despite the crude approximation of noninteracting electrons, this resulting energy formula has the correct structure and even the numerical factors are not ridiculously off target. For, if one does a full-blown realistic calculation, the outcome is

$$\frac{-E}{\frac{1}{2}Z^2e^2/a_0} = 1.537Z^{1/3} - 1 + 0.540Z^{-1/3}, \qquad (6.5.14)$$

a remarkably simple expression for the total binding energy of neutral many-electron atoms.

The leading term in (6.5.14) is known as the *Thomas–Fermi energy* and accounts for the bulk electrostatic interaction (named after Llewellyn H. Thomas and Enrico Fermi). The next term, the simple -1, is the so-called *Scott correction* (J. M. C. Scott) which results from a better treatment of the innermost, the most strongly bound electrons. For them, the singularity of the Coulomb potential of the nuclear charge is quite important. Finally, the third term, derived by Julian Schwinger, accounts for the exchange energy among the electrons (nine eleventh) and a correction to the kinetic energy beyond its Thomas–Fermi approximation (two eleventh).

Note that there is no sign of atomic shells in the energy formula (6.5.14). The shells manifest themselves in the next, rather complicated, term that is oscillatory with an amplitude proportional to $Z^{2/3}$ and a period proportional to $Z^{1/3}$. The ubiquitous dependence on $Z^{1/3}$ originates in $n_s \sim N^{1/3} = Z^{1/3}$ as we established in (6.5.9).

This oscillatory term of the shell-filling represents contributions from the outermost electrons, which are so loosely bound that they cannot possibly contribute substantially to the total binding energy. We expect, therefore, that (6.5.14) is a very good approximation. Indeed it is.

To demonstrate the case, we compare the approximate formula (6.5.14) with the exact binding energies in this figure:

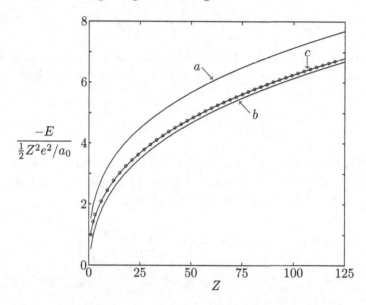

The circles indicate the exact binding energies for $Z = 1, 2, 3, 6, 9, \ldots, 120$ and the smooth curves represent the successive approximations of (6.5.14),

$$\frac{-E}{\frac{1}{2}Z^2 e^2 / a_0} = \begin{cases} \text{curve } a: & 1.537 Z^{1/3}\,, \\ \text{curve } b: & 1.537 Z^{1/3} - 1\,, \\ \text{curve } c: & 1.537 Z^{1/3} - 1 + 0.540 Z^{1/3}\,. \end{cases} \qquad (6.5.15)$$

Curve c goes right through the circles, except for hydrogen ($Z = 1$), where one would not expect the approximation to work in the first place.

What can we say about the size of many-electron atoms? Whereas a detailed answer would involve much more machinery than we have at our disposal, we can give a good rough answer. For this purpose, we note that the Z dependence of the Hamilton operator

$$H = \sum_{j=1}^{N} \left(\frac{1}{2M} \vec{P}_j^2 - \frac{Z e^2}{|\vec{R}_j|} \right) - \frac{1}{2} \sum_{j \neq k = 1}^{N} \frac{e^2}{|\vec{R}_j - \vec{R}_k|} \qquad (6.5.16)$$

is solely in the electron-nucleus interaction term,

$$\frac{\partial H}{\partial Z} = -e^2 \sum_{j=1}^{N} \frac{1}{|\vec{R}_j|} .$$ (6.5.17)

According to the Hellmann–Feynman theorem, we thus have

$$\left\langle \sum_j |\vec{R}_j|^{-1} \right\rangle = -\frac{1}{e^2}\left\langle \frac{\partial H}{\partial Z} \right\rangle = \frac{1}{e^2}\frac{\partial}{\partial Z}(-E)$$ (6.5.18)

where the binding energy $-E$ is to be regarded as a function of Z and N. The leading term, as we found it in (6.5.12), is $-E \propto (e^2/a_0)Z^2 N^{1/3}$, so that

$$\left\langle \sum_j |\vec{R}_j|^{-1} \right\rangle \propto \frac{1}{a_0} Z N^{1/3} .$$ (6.5.19)

On the left we have the sum of inverse distances of all electrons. We divide by N to get the mean inverse distance for an electron, and take the reciprocal to get the mean distance itself. This yields

$$\text{mean electron distance} = \left(\frac{1}{N}\left\langle \sum_j |\vec{R}_j|^{-1} \right\rangle \right)^{-1}$$
$$\propto a_0 N^{2/3}/Z \Big|_{N \,=\, Z} = a_0/Z^{1/3} .$$ (6.5.20)

So, a neutral atom with nuclear charge Z has a size given by (a numerical multiple of) the Bohr radius a_0 divided by $Z^{1/3}$. Somewhat surprisingly, we therefore learn that atoms with many electrons are of smaller size than atoms with few electrons. This is a consequence of the nuclear attraction dominating over the electron-electron repulsion, which has a tendency of averaging out to some extent because a single electron experiences repulsive forces from different directions, while all electrons are jointly attracted toward the nucleus.

It should be kept in mind that these observations concern the average distance of electrons from the nucleus. A different question, and one that is more relevant for chemistry, would ask about the distance from the nucleus of the most weakly bound electrons, those participating in spectroscopy and in the electron redistribution in chemical reactions. These outermost electrons are at a distance from the nucleus that is essentially independent of the atomic number $N = Z$.

Index

Note: Page numbers preceded by the letters BM or SS refer to
Basic Matters and *Simple Systems*, respectively.